SUSTAINABILITY AND ENERGY MANAGEMENT FOR WATER RESOURCE RECOVERY FACILITIES

WEF Manual of Practice No. 38
ASCE Manuals and Reports on Engineering Practice No. 137

2018

Water Environment Federation
601 Wythe Street
Alexandria, VA 22314-1994 USA
http://www.wef.org

American Society of Civil Engineers/
Environmental and Water Resources
Institute
1801 Alexander Bell Drive
Reston, VA 20191-4400
http://www.asce.org

ISBN: 978-1-57278-341-6

Water Environment Research, WEF, and *WEFTEC* are registered trademarks of the Water Environment Federation. American Society of Civil Engineers, ASCE, Environmental and Water Resources Institute, and EWRI are registered trademarks of the American Society of Civil Engineers.

About WEF

The Water Environment Federation (WEF) is a not-for-profit technical and educational organization of 33,000 individual members and 75 affiliated Member Associations representing water quality professionals around the world. Since 1928, WEF and its members have protected public health and the environment. As a global water sector leader, our mission is to connect water professionals; enrich the expertise of water professionals; increase the awareness of the impact and value of water; and provide a platform for water sector innovation. To learn more, visit www.wef.org.

About ASCE/EWRI

Founded in 1852, the American Society of Civil Engineers (ASCE) represents more than 150,000 members of the civil engineering profession worldwide and is America's oldest national engineering society. Created in 1999, the Environmental and Water Resources Institute (EWRI) is an Institute of ASCE. EWRI services are designed to complement ASCE's traditional civil engineering base and to attract new categories of members (non-civil engineer allied professionals) who seek to enhance their professional and technical development.

For information on membership, publications, and conferences, contact

ASCE/EWRI
1801 Alexander Bell Drive
Reston, VA 20191-4382
(703) 295-6000
http://www.asce.org

Prepared by **Design of Water Resource Recovery Facilities Task Force** of the **Water Environment Federation** and the **American Society of Civil Engineers/ Environmental and Water Resources Institute**

Terry L. Krause, P.E., BCEE,
    WEF Fellow, *Chair*

Jeanette Brown, P.E., DEE,
    D. WRE; Mark E. Lang,
    P.E., BCEE; and Kendra
    D. Sveum, P.E., *Volume
    Leaders*

Jennifer L. Strehler, P.E., MBA,
    BCEE, ENV-SP;
    Rebecca Gauff; Elizabeth
    Miner; Jason Turgeon;
    and David Valero, *Authors*

Malarmagal Ahilan
Rafid Alkhaddar (ASCE Blue
    Ribbon Review Panel)
Naomi Eva Anderson
Rich Atoulikian
Eric Auerbach
Jeovanni Ayala-Lugo, P.E.
Amber Batson, P.E.
Paul Bizier, P.E., F.ASCE,
    D.WRE
Lucas Botero, P.E., BCEE,
    ENV SP
Akram Botrous, Ph.D., P.E.,
    BCEE
Lisa Boudeman
Keith Bourgeous, Ph.D, P.E.
Gregory A. Bowden
John R. Bratby, Ph.D., P.E.
Kari Fitzmorris Brisolara, ScD,
    MSPH, QEP
John P. Brito, P.E.
Lewis Bryant, P.E.
Marie S. Burbano, Ph.D., P.E.,
    BCEE

Misti Burkman
Chris Bye
Jennifer Callahan
Onder Caliskaner, Ph.D., P.E.
Leonard W. Casson, Ph.D., P.E.,
    BCEE
Stan Chilson
S. Rao Chitikela, Ph.D., P.E.,
    P.Eng, BCEE (ASCE Blue
    Ribbon Review Panel)
Timothy A. Constantine
Bruce L. Cooley, P.E.
Emma Cooney
Chris D. Cox, Ph.D., P.E.
    (ASCE Blue Ribbon Review
    Panel)
Glen T. Daigger, Ph.D., P.E.,
    BCEE, NAE (ASCE Blue
    Ribbon Review Panel)
Chris DeBarbadillo, P.E.
Adam Dellinger
Timur Deniz, Ph.D., P.E., BCEE
Carlos Diaz
Bruce DiFrancisco, P.E.
Ludwig Dinkloh
Alexandra Doody, P.E.
Leon Downing
Bertrand Dussert
Na-Asia Ellis
Adam Evans, P.E.
Richard Finger
William Flores
Kristin Frederickson
Daniel Freedman, P.E.
Val S. Frenkel, Ph.D., P.E.,
    D. WRE
John Friel, P.E.
Edward W. Fritz
Rebecca Gauff

A. Robert Rubin, Ph.D.
Andrew Salveson, P.E.
Domenico Santoro
Patricia A. Scanlan
Kimberly Schlauch
Harold E. Schmidt, Jr., P.E.,
    BCEE
Kenneth Schnaars, P.E.
Megan Yoo Schneider
Sandra Schuler
Matt Seib, Ph.D.
Douglas Sherman, P.E.
Toshio Shimada, Ph.D., P.E.
Jim E. Smith, Jr, D.Sc, MASCE,
    BCEEM
Eric Spargimino, P.E., LEED AP
Eric T. Staunton, Ph.D.
Jennifer L. Strehler, P.E., MBA,
    BCEE, ENV-SP
Timothy H. Sullivan, P.E.
Alex Szerwinski, P.E.
Steven Swanback
Jay L. Swift, P.E.
Alex Tabb
Berrin Tansel, Ph.D., P.E.,
    BCEE, FASCE, FEWRI
    (ASCE Blue Ribbon
    Review Panel)
Anthony Tartaglione, P.E.,
    BCEE
George Tchobanoglous (ASCE
    Blue Ribbon Review Panel)

Matt Tebow, P.E.
Rachelle Tippetts
David Tomowich
K. Richard Tsang , Ph.D., P.E.,
    BCEE
Jason Turgeon
Andrea Turriciano White, P.E.
David Ubert
David Valero
Don Vandertulip, P.E., BCEE
Ales Volcansek, P.E.
Tanush Wadhawan, Ph.D.
Trevor Wagenmaker, P.E.
Kristen Waksman
Steve Waters, P.E., P. Eng
David G. Weissbrodt, Asst.
    Prof., Ph.D., M.Sc.,
    Dipl.-Ing.
Jianfeng Wen
Curt Wendt, P.E., CAP
Claes Westring
Jason J. Williams, P.E.
Matthew J. Williams, P.E.
Hannah T. Wilner, P.E., PMP
Melissa K. Woo, P.E.
Paul Wood
Wade Wood, P.E.
Thomas Worley-Morse, Ph.D.
Usama Zaher, Ph.D., P.E.
Tian C. Zhang, Ph.D., P.E.,
    BCEE, F.ASCE (ASCE Blue
    Ribbon Review Panel)

Under the Direction of the **Municipal Subcommittee** of the **Technical Practice Committee**

2018

Water Environment Federation
601 Wythe Street
Alexandria, VA 22314-1994 USA
http://www.wef.org

Contents

List of Tables

List of Figures

Authors' and reviewers' efforts were supported by the following organizations:

AECOM, Piscataway, New Jersey; Buffalo, New York

American Water, Voorhees, New Jersey

Arcadis U.S., Inc., Highlands Ranch, Colorado; Buffalo, New York; White Plains, New York

Arvos Schmidtsche Schack LLC, Wexford, Pennsylvania

Automation Federation, Raleigh, North Carolina

Barge, Waggoner, Sumner and Cannon, Nashville, Tennessee

Bedrock Enterprises, Inc., Baden, Pennsylvania

Black & Veatch, Coral Springs, Florida; Indianapolis, Indiana; Overland Park, Kansas; Kansas City, Missouri; St. Louis, Missouri; Memphis, Tennessee

Brown and Caldwell, Maitland, Florida; Orlando, Florida; Charlotte, North Carolina; Nashville, Tennessee; Alexandria, Virginia; Seattle, Washington

Carollo Engineers, Costa Mesa, California; Walnut Creek, California; Littleton, Colorado; Tampa, Florida; Dallas, Texas

CDM Smith, Carlsbad, California; Irvine, California; Los Angeles, California; Denver, Colorado; Bogota, Columbia; Maitland, Florida; Miami, Florida; Orlando, Florida; Boston, Massachusetts; Manchester, New Hampshire; Albany, New York; Raleigh, North Carolina; Providence, Rhode Island; Houston, Texas; Austin, Texas; Dallas, Texas; Fairfax, Virginia; Leesburg, Virginia; Bellevue, Washington

CH2M, Tampa, Florida; Chicago, Illinois; Albuquerque, New Mexico; Herndon, Virginia; Toronto, Ontario, Canada

Corrosion Probe, Inc., Centerbrook, Connecticut

DC Water, Washington, D.C.

Donohue & Associates, Inc, Chicago, Illinois

Dynamita, Toronto, Canada

Dynamita S.A.R.L., Nyons, France

EnviroSim Associates Ltd., Hamilton, Ontario, Canada

Evoqua Water Technologies LLC, Bradenton, Florida

Garver, Dallas, Texas; Frisco, Texas

Gray and Osborne, Seattle, Washington

GREELEY and HANSEN, Chicago, Illinois, San Francisco, California

Hazen and Sawyer, Raleigh, North Carolina

HDR Engineering, Inc., Walnut Creek, California; Calverton, Maryland; Cleveland, Ohio; Nashville, Tennessee

Hubbell, Roth & Clark, Inc., Detroit, Michigan

inCTRL Solutions Inc., Oakville, Ontario, Canada
Intera, Richland, Washington
Johnson County Wastewater, Olathe, Kansas
Kennedy/Jenks Consultants, San Francisco, California
Kimley-Horn and Associates, Inc., Mesa, Arizona; Ocala, Florida;
    Tampa, Florida; West Palm Beach, Florida; Ft. Worth, Texas
Laura Marcolini & Associates, Inc., Cumberland, Rhode Island
Louisiana State University, Baton Rouge, Louisiana
Madison Metropolitan Sewerage District, Madison, Wisconsin
Manhattan College, Bronx, New York
Material Matters, Elizabethtown, Pennsylvania
National Automation, Inc., Spring, Texas
NOLASCO y Asociados. S. A., Buenos Aires, Argentina
North Carolina State University, Raleigh, North Carolina
SafeStart, Belleville, Ontario, Canada
Short Elliott Hendrickson Inc., St. Paul, Minnesota
Smith and Loveless, Inc., Lenexa, Kansas
Southeast Environmental Engineering, LLC, Knoxville, Tennessee
St. Croix Sensory, Inc., Stillwater, Minnesota
Stantec Consulting Services, Rocklin, California; Denver, Colorado;
    Tampa, Florida; Portland Oregon
Tesco Controls, Inc., Sacramento, California
Total Safety Compliance, Mesa, Arizona
University of Pittsburgh, Pittsburgh, Pennsylvania
URS Corporation, Buffalo, New York
U.S. Environmental Protection Agency, Boston, Massachusetts
V&A Consulting Engineers, Houston, Texas
Vandertulip WateReusEngineers, San Antonio, Texas
Varec Biogas, Stafford, Texas
Veolia North America, Chicago, Illinois
Washington State Department of Ecology, Bellevue, Washington
WesTech Engineering, Salt Lake City, Utah
Xylem Inc., White Plains, New York

Sustainability and Energy Management for Water Resource Recovery Facilities

Jennifer L. Strehler, P.E., MBA, BCEE, ENV-SP; Rebecca Gauff; Elizabeth Miner; Jason Turgeon; and David Valero

1.0 INTRODUCTION

Energy savings and sustainable design deserve special attention to be sure water resource recovery facilities (WRRFs) have long-term adaptability and resilience to global climate change, volatile energy prices, and other predictable change scenarios. Municipal WRRFs in the United States use approximately 30.2 bil. kWh/yr, or approximately 0.8% of total electricity use in the United States (EPRI, 2013). Yet, of the approximately 14,780 WRRFs in the United States, only approximately 1268 (8.4%) include anaerobic digestion (which offers the potential to recovery chemical energy) and beneficially use this energy on site for production of power and/or heat (WEF, 2013).

The umbrella of *sustainability* covers long-term provisions for resilient facilities to manage a wider range of stressors, and treatment process adaptability to accommodate changing regulations. Sustainability in this context refers to the ability to continue operating without causing immediate or long-term harm to the environment, society, or depleting natural resources. In the accounting sense, this means planning for the future by making annual financial investments that seek to minimize the total life cycle cost of a WRRF across its full life and avoid deferring costs and negative effects to future generations. The concept of sustainability has also expanded to include indirect effects to the greater community, and consider local industry partnerships and social justice issues. Optimizing the sustainability of a WRRF requires

1

systems thinking and a willingness to make change. Using strong leadership, cultural buy-in, and an understanding of the best technical practices described in this publication, utility managers can make a big difference.

Public expectations of today's wastewater utility have changed in recent years and are likely to continue to evolve. There is greater emphasis on making appropriate long-term investments for the future and a lower overall *life cycle cost*. This includes an increased expectation to prioritize sustainability of operations and maintenance. These changes are congruent with a shift toward resource recovery and the expectation for more readily realizable benefits to be derived from WRRF projects.

The focus now is on planning and designing facilities that are less costly to operate and maintain over the facility's full life cycle and more resilient to near-term and long-term change. New facilities are expected to have significantly reduced energy consumption, modern automated equipment with less downtime, cost less to operate and maintain, be more resilient to weather extremes, require less chemical inputs, and be able to offer marketable products through resource recovery. These trends are driving facility managers to better balance utility finances through improved control over operating energy and chemical use, and through creation of new revenue sources.

We are also witnessing an increased public expectation for wastewater utilities to establish defined goals that reduce negative aspects and improve the benefits of wastewater infrastructure for future generations. Accordingly, the public expects greater community engagement in planning public infrastructure projects, more attention to *equity and social justice* with respect to facility siting and other neighborhood effects, inclusion of public amenities, and overall more transparency and justification in decisions made by local government. Leadership, vision, transparency, long-term view, fairness, and collaboration are skills needed in this arena. Now is the time for utilities managers to relish the opportunity to engage with the public and spotlight the important work done at WRRFs.

When commencing an initial review of sustainability and/or energy management for WRRFs, the first step is to establish the boundaries for the analysis. This includes setting economic parameters, environmental analyses, and/or geographic/community boundaries. Boundaries will be influenced by the context and culture of the organization, typical approaches used for decision-making, and the current stage of the project. For example, will the evaluation be limited to the immediate project site or a larger campus managed by the same owner? Is the owner interested in applying external frameworks to seek awards (e.g., Leadership in Energy Environmental Design [LEED], Envision, BREEAM®), obtain grant/loan funds, or to simply facilitate communication? Context and boundary decisions dictate final outcomes so it is important to do this early and get input from decision-makers.

This publication addresses energy and sustainability of operations at WRRFs, with a focus on facilities in the United States and Canada but includes strategies, tools, and references that are globally applicable. The best practices described here draw from examples in which optimization of constrained resources within a defined system has achieved the best set of long-term benefits. The spatial boundaries are generally the property line and/or the immediately affected neighborhood near the WRRF. However, indirect effects of facility operations are also discussed regarding the embodied energy of construction materials and greenhouse gases associated with these materials and imports/exports of the facility. *Embodied energy* is the non-renewable energy consumed by all of the processes associated with the production of a building, from the mining and processing of natural resources to manufacturing, transport, and product delivery to the construction site; measured in gigajoules per unit weight or per unit area. *Greenhouse gases* (GHGs) include the collection of gases both naturally occurring and human-made (e.g., carbon dioxide, methane, nitrous oxide, ozone, chlorofluorocarbons, and other compounds), which absorb infrared radiation and contribute to the atmospheric greenhouse effect.

Regarding service life, evaluation of energy and sustainability spans the full expected life of a WRRF, which may range from 30 to 100 years or longer; with 20 to 30 years typically used as the horizon for alternatives analysis and facility planning. Considerations across the project life (i.e., during planning, design, construction, bidding, and operation) are described herein. Guidance for improving energy and sustainability at WRRFs is constantly evolving. Please also see the "References" and "Suggested Readings" sections at the end of this publication.

2.0 SUSTAINABILITY FRAMEWORKS

A number of frameworks for integrating sustainability concepts into projects have been developed in the last two decades. Table 1 provides a summary of sustainability frameworks and related references relevant to the wastewater industry. The choice of which framework is most appropriate for a given project depends on the overall project goals, geographic location, intended use of the installed facility, and owner preferences. Regardless of which tool is used to help integrate sustainability into a project, the project team should anticipate that there will be tradeoffs between competing objectives during each phase and plan ahead as to how objectives will be prioritized and the resulting facility optimized. These tradeoffs are themselves a function of previous decisions about economic, environmental, and community boundaries.

TABLE 1   Summary of sustainability frameworks.

Framework	Intended applications	References
BREEAM	New pilot (2016) international planning, design, and construction of infrastructure	http://www.breeam.com/infrastructure (BRE, 2017)
CRWU	Planning and assessment for adaptation to climate change	https://www.epa.gov/crwu (U.S. EPA, 2017)
ENVISION	Planning, design, and construction of civil infrastructure	https://sustainableinfrastructure.org/ (ISI, 2017)
LEED	Planning, design, construction, and operation of green buildings	http://www.usgbc.org/LEED/ (USGBC, 2017)
ASTM	Sustainability standards applicable across a broad range of industries	http://www.astm.org/COMMIT/sustain.html (ASTM International, 2017a)
ISO 14001	Environmental management—Requirements and guidance for use	https://www.iso.org/iso-14001-environmental-management.html (ISO, 2015)
ISO 14040	Life cycle assessments—Principles and framework	https://www.iso.org/standard/37456.html (ISO, 2006a)
ISO 14044	Life cycle assessments—Requirements and guidelines	(ISO, 2006b)
ISO 50001	Energy management—Requirements and guidance for use	http://www.iso.org/iso/home/standards/management-standards/iso50001.htm (ISO, 2011)

Sustainability and energy management frameworks in widespread use and/or that are specifically designed for WRRFs are summarized here. Some resiliency decision support tools focus on practical issues such as planning for short-term power outages, while others encourage big-picture thinking such as preparing for rising sea levels or changing precipitation patterns. These tools can be used as standalone resources, or grouped together as needed. New tools and methodologies are continually being introduced and refined. Engineers need to pay close attention to this fast-moving field to stay abreast of the most current best practices.

2.1   U.S. Environmental Protection Agency Climate Ready Water Utilities Initiative

Understanding and adapting to climate change threats is an important part of decision-making to improve *resilience* for water, wastewater, and stormwater utilities. Extreme events like floods, drought, sea-level rise, wildfires, and reduced snowpack may become more frequent or more intense. Planning for extreme events can help protect utility infrastructure and operations, allowing utilities to provide reliable and sustainable service to their customers.

U.S. EPA's Climate Ready Water Utilities (CRWU) initiative helps utilities actively plan for climate change effects. The CRWU initiative provides tools, training, and technical assistance to adapt to climate change by promoting a clear understanding of climate science and adaptation options. This preparation involves developing an understanding of the risks that result from changes in climate, planning to address these effects, and implementing adaptive actions to reduce the consequences of climate change. The CRWU's approach is an iterative process that includes updating actions as new climate data and tools become available to support continuous improvement. Available tools and resources include

- Climate Resilience Evaluation and Awareness Tool—This free online tool is a climate risk assessment with planning application for water sector utilities. This tool helps users discover which extreme weather events pose significant challenges and build scenarios to identify potential effects, assess critical assets and the actions to protect them from the consequences of climate change on utility operations, and generate reports to share information about the costs and benefits of risk reduction strategies for decision-makers and shareholders.

- Workshop planning—The *Workshop Planner for Climate Change and Extreme Events Adaptation* presents a four-step process to develop and conduct climate change adaptation workshops. Tailored materials are generated by this tool for each utility's specific workshop needs.

- Adaptive strategies—The *Adaptation Strategies Guide for Water Utilities* is a document with examples of water utility impacts and adaptation options (U.S. EPA, 2015). The guide includes sustainability briefs to support adaptation-planning, examples of utilities implementing adaptation options, and worksheets to help with the planning process. As a companion to the guide, a number of stand-alone utility case studies are also available.

- Additional resources—The CRWU Initiative website (www.epa.gov/crwu) maintains a list of resources and offers periodic webinars.

2.2   Envision

Envision is a framework developed and administered by the Institute for Sustainability (ISI). Unlike LEED (which is intended for vertical construction and occupied buildings), Envision is specifically intended for unoccupied public infrastructure, including WRRFs. The Envision framework includes 60 criteria spanning five major areas as follows:

- Quality of life—purpose, well-being, community;
- Leadership—collaboration, management, planning;
- Resource allocation—materials, energy, water;
- Natural world—siting, land and water, biodiversity; and
- Climate and risk—emissions, resilience.

The project team selects which criteria are applicable to the project and exempts others; this step determines the total achievable score. The level of achievement under each criterion determines the project score and is described across the following range:

- Conventional—state of practice (lowest level);
- Improved—encouraging sustainability;
- Enhanced—"on the right track";
- Superior—remarkable performance;
- Conserving—"zero negative impacts"; and
- Restorative—restoring natural resources, ecological, economic, and social systems to more natural or predevelopment state (highest level).

The Envision framework can be accessed and used by anyone at no cost, regardless of intent to apply for an award. Fees are charged by ISI to projects seeking Envision awards. Award-seeking projects require an ENV-SP certified professional to prepare the award application and work with appointed verifier and authenticator to complete the assessment process. Application for award is typically made at approximately 90% engineering design level. Award levels include bronze, silver, gold, and platinum.

2.3   Leadership in Energy and Environmental Design

The LEED framework is a third-party certification for the design, construction, and operation of high-performance green buildings. This framework was developed by and is administered by the U.S. Green Building Council (USGBC). Leadership in Energy and Environmental Design was created to encourage and accelerate the global adoption of sustainable green buildings and to develop practices through the creation and implementation of universally understood and accepted tools and performance criteria. The current version is LEED v4 (USGBC, 2016). The LEED framework has rating systems for several project types: building design and construction (BD+C); building operations and maintenance (O&M); interior design and construction; homes; and neighborhood development.

The LEED framework applies primarily to occupied buildings and is not intended to apply to unoccupied/semi-occupied buildings such as pump stations, or to processes, such as WRRFs or tanks. This is because many LEED credits are based on design decisions that affect building occupants. However, specific occupied buildings within WRRFs, such as an administration building or a maintenance building, can be eligible for LEED BD+C and/or LEED O&M certification. Under the LEED framework, buildings that meet the established criteria can be awarded the following designations: certified, silver, gold, or platinum. These LEED rating systems focus sustainable design efforts in seven key areas, referred to as credit categories, which include both mandatory prerequisites and optional credits:

- Location and transportation,
- Sustainable sites,
- Water efficiency,
- Energy and atmosphere,
- Material resources,
- Indoor environmental quality, and
- Innovation and regional priority.

The LEED framework can be used for design guidance at no cost in a limited manner, regardless of intent to pursue award in a limited manner; some LEED tools are only available to officially registered projects. Projects must meet minimum program requirements to be eligible for LEED certification and award. Projects seeking LEED certification should consult the USGBC website early in the design process to understand LEED project registration, application requirements, and timelines. After successfully completing project registration, the application for award timeline depends on which LEED rating system is being used since performance data may be required. The Green Business Certification Inc.

administers LEED by performing third-party review and verification of LEED registered projects.

2.4   American Society for Testing and Materials

The American Society for Testing and Materials (ASTM) develops consensus standards and provides a list of sustainability standards applicable to many industries. Table 2 lists ASTM standards that are most likely to be applicable to the design of a WRRF. The ASTM maintains a database referencing over 500 ASTM standards plus more than 300 other standards and programs from organizations involved in sustainability. The database is maintained by ASTM Committee E60 on Sustainability. Contact www.ASTM.org for a complete list of standards.

TABLE 2   Summary of relevant ASTM standards.

Standard	Intended applications	Reference
ASTM E2114	Buildings and Construction—Standard Terminology for Sustainability Relative to the Performance of Buildings	https://www.astm.org/Standards/E2114.htm
ASTM E2129	Buildings and Construction—Standard Practice for Data Collection for Sustainability Assessment of Building Products	https://www.astm.org/Standards/E2129.htm
ASTM E2432	Buildings and Construction—Standard Guide for General Principles of Sustainability Relative to Buildings	https://www.astm.org/Standards/E2432.htm
ASTM E2921	Buildings and Construction—Standard Practice for Minimum Criteria for Comparing Whole Building Life Cycle Assessments for Use with Building Codes, Standards, and Rating Systems	https://www.astm.org/Standards/E2921.htm
ASTM E2635	Water Use and Conservation—Standard Practice for Water Conservation in Buildings Through In-Situ Water Reclamation	https://www.astm.org/Standards/E2635.htm
ASTM E2717	Water Use and Conservation—Standard Practice for Estimating the Environmental Load of Residential Wastewater	https://www.astm.org/Standards/E2717.htm
ASTM E2727	Water Use and Conservation—Standard Practice for Assessment of Rainwater Quality	https://www.astm.org/Standards/E2727.htm
ASTM E2728	Water Use and Conservation—Standard Guide for Water Stewardship in the Design, Construction, and Operation of Buildings	https://www.astm.org/Standards/E2728.htm

2.5   Customized Frameworks for Environmental and Energy Management Systems

Any organization can develop a customized Environmental Management System (EMS) framework to help it achieve its environmental goals through consistent review, evaluation, and improvement of its environmental performance in a business-specific manner. Customized EMS plans are tailored to the utility's individual objectives and targets for environmental performance. This allows the organization to address specific regulatory demands and unique challenges in a systematic and cost-effective manner. This is a proactive approach that can reduce the risk of non-compliance while also improving health and safety practices for employees and the public. An EMS can also help address non-regulated issues, such as energy conservation, and can promote stronger operational control and employee stewardship. The core assumption in an EMS is that consistent review and evaluation will identify opportunities for improving and implementing the environmental performance of the organization; thus, the organization's leadership must openly welcome feedback from employees. Basic elements of an EMS include

- Reviewing the organization's environmental goals;
- Analyzing its environmental impacts and legal requirements;
- Setting environmental objectives and targets to reduce environmental effects and comply with legal requirements;
- Establishing programs to meet these objectives and targets;
- Monitoring and measuring progress in achieving the objectives;
- Ensuring employees' environmental awareness and competence; and
- Reviewing progress of the EMS and making improvements.

The International Organization for Standardization (ISO) has developed several standards for sustainability based on the EMS framework and a plan of continuous improvement. ISO 14001 and ISO 50001 are described here.

2.5.1   ISO 14001: Environmental Management Systems

The most commonly used framework for an EMS is the ISO 14001 standard. Established in 1996, this framework is the official international standard for an EMS and is based on the Plan-Do-Check-Act methodology. Figure 1 summarizes the ISO 14001 approach toward continuous improvement in environmental management.

FIGURE 1 Environmental management system cycle of continuous improvement.

The five main stages of an EMS, as defined by the ISO 14001 standard, are

1. Commitment and Policy—Top management commits to environmental improvement and establishes the organization's environmental policy as a foundational step;

2. Planning—Environmental aspects of operations, such as air pollutants or hazardous waste, are identified. Top organizational objectives—such as worker health and safety, environmental compliance, cost, and so forth—are identified. An objective is an overall environmental goal (e.g., minimize use of chemical X). Targets that are detailed and quantifiable arising from each objective are set (e.g., reduce use of chemical X by 25% by January 1, 2030). The final part of the planning stage is devising an action plan for meeting the targets. This includes designating responsibilities, establishing a schedule, and outlining defined active steps to meet targets;

3. Implementation—The organization follows through with the action plan using the necessary resources (human, financial, etc.). An important component is employee training and awareness for all employees. Other steps include documentation, following operating procedures, and setting up internal and external communication lines;

4. Evaluation—The utility monitors its operations to evaluate whether targets are being met. An objective third party may be used for the evaluate step; and

5. Review—Top management reviews the results of the evaluation to see what parts of the EMS are working and what may need to be adjusted or re-vamped. Management should also verify that original environmental policy is still consistent with the organization's current values or adjust it accordingly. If targets have not been met but are still viewed as valid, then corrective action is taken and/or the EMS is updated. The review stage creates a loop of continuous improvement.

In 2000, the Lowell Regional Wastewater Utility (LRWWU) became the first public entity in the United States to certify all of its divisions to the ISO 14001 Standard. The facility decided to implement its EMS through the *1st EMS Initiative for Local Governments 1997–1999*, a U.S. EPA-supported program facilitated by the Global Environment & Technology Foundation (www.getf.org) (U.S. EPA, 2005). The city has maintained certification since 2000, instituting an incentive program whereby employees are given financial rewards for successful third-party audits and maintaining certification. This LRWWU facility has reduced energy use by more than 35% since 1995, with total savings estimated at more than $3 million. Projects implemented under the EMS include pump; blower; lighting; and building heating, ventilation, and air conditioning (HVAC) efficiency upgrades; rooftop solar; and solar thermal heating (Young, 2016).

2.5.2 ISO 50001: Energy Management Systems

ISO released the ISO 50001 standard in 2011. This standard is based on the same principles of continuous improvement as the 14001 EMS, but is specifically focused on energy. The U.S. Department of Energy (DOE) is facilitating application of ISO 50001 by partnering with water and wastewater treatment agencies via a program called "Superior Energy Performance" or the SEP™ program. Program goals are to inspire continual improvement in WRRFs, reduce operating costs, reduce GHG emissions, and lower demand for energy.

The DOE provides guidance, tools, and energy management experts to train and assist participants through this formal program. As of the date of this publication, seven WRRFs—referred to as "Better Plants partners"—were participating in the SEP™ Water and Wastewater Pilot Project. Participants integrate energy management in their business operations and culture through a systematic approach to managing energy. Partners agree to implement specific improvements and strive to meet SEP™ requirements at one of their WRRFs and evaluate combined heat and power (CHP) as an option for those with anaerobic digesters. Participants also agree to support third-party validation and audit steps to verify that the program requirements were met and to collect and provide data back to DOE to support knowledge transfer. Successful implementation results in issuance of SEP™ and ISO 50001 certificates, as well as national recognition. More information is available at http://energy.gov/eere/amo/better-plants.

2.6   Projects Seeking Awards Under Multiple Frameworks

If the project team is pursuing award certification under one or more frameworks, it is important to establish a budget for documentation early in the project. It is best to appoint an accredited professional to be the sustainability lead who facilitates the integrated design process and uses tracking tools to verify all required elements of sustainability are integrated into the project. Additional planning and management time should be expected if the project team intends to seek award under multiple frameworks. Be prepared for areas of overlap but with differing evaluation criteria and documentation requirements. For example, both LEED and Envision frameworks require application of *triple bottom line* decision-making, a significant level of public engagement, and require detailed calculations for energy and water conservation, embodied energy, and renewable energy. However, LEED calculates on-site energy production ratios on the basis of utility cost (e.g., dollars per year) using ASHRAE 90.1-G modeling, whereas Envision calculates this on the basis of energy units (e.g., megajoules per year) and allows a variety of modeling approaches. Nevertheless, both frameworks can be successfully applied to a larger WRRF project. It is helpful to contact the certifying agencies early to let them know of the intent to apply multiple frameworks and to determine if different site boundaries should be set within the project to support the intended application of each framework (e.g., LEED boundary around the administration building; Envision boundary around entire facility site).

3.0 PRACTICAL APPLICATION

Practical application of the sustainability and energy management happens when key decisions are made during the planning, preliminary design, final design, construction, and operational phases of a WRRF capital project. This section highlights key issues and considerations for imparting greater sustainability across the life cycle of a typical WRRF project.

3.1 Planning Phase

During the planning phase the owner, project team, and stakeholders typically have the greatest influence over the ultimate sustainability of the resulting facility. This phase includes goal setting, siting decisions, technology selection, and alternatives screening and evaluation. The alternatives selection process typically follows a process as shown in Figure 2.

3.1.1 Goal Setting

Every project is different and every utility and community has a different set of expectations about the project outcomes. For this reason, a unique set of project goals

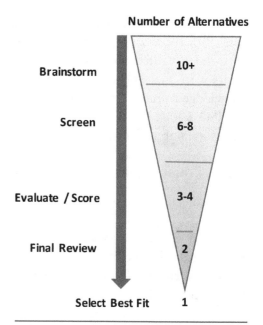

FIGURE 2 Alternatives development, evaluation, and selection process.

and objectives should be developed early. This is a high-level activity that also establishes the critical success factors of the project and realistic constraints. This information is needed to support subsequent steps to screen and evaluate technical alternative solutions.

3.1.1.1   Goal Setting Tools

Local policies, benchmarking, gap-analysis approaches, energy audits, and eco-charrettes are useful sources of information and tools in the goal-setting phase. *Eco-charrettes* are collaborative meetings conducted in the early design stage to brainstorm and establish sustainability development goals, strategies, and integrated design solutions and which include participants involved in design, construction, and operation.

3.1.1.1.1   POLICY DIRECTIVES

Sustainability and energy management goals and objectives for a given project or capital program should be clearly linked to existing local policies and community plans for energy conservation, GHG reduction, renewable energy, recycling, water conservation, and other measures of sustainability. Defining these links to policy directives should be a specific and early-out task in a design engineer's contract. To this end, owners can include the desired level of experience with sustainability and energy management as key qualifications in consultant request for proposals (RFPs) for facility planning and design contacts. The Consortium for Energy Efficiency (CEE) offers guidance on issuing RFP/RFQ documents that seek to include for energy efficiency elements into new or refurbished WRRFs (CEE, 2010).

3.1.1.1.2   BENCHMARKING AND GAP ANALYSIS

Benchmarking and gap analysis are useful tools to assess the utility's starting position, understand performance relative to internal goals or other similar utilities, and to assist leaders to craft a vision of the sustainable utility of the future. If you have not yet benchmarked your operation from an energy and sustainability perspective, now is the time. Benchmarking is the first step in assessing opportunity for improvement and answers questions such as: "How does my utility compare to 'industry average' or to 'best practices' in the industry?" and "In what areas could we do better?" Measuring and managing the energy consumption and *carbon footprint* (the cumulative total amount of carbon dioxide and other carbon compounds emitted to the atmosphere as a result of combustion of non-renewable fossil fuels consumed by the ongoing operation of the WRRF) of a WRRF can motivate employees to do business with greener suppliers, improve the energy efficiency of equipment, reduce waste in operations, and to re-examine disposal options and resource recovery opportunities because they see sustainability is being measured and valued. Benchmarking and gap analysis consists of the following steps:

1. What is the "industry average" level of performance for a similar facility?
   - Define what performance metrics to compare. For example, O&M cost, natural gas consumption, or on-site electrical power production;
   - Seek out operating data from facilities with similar characteristics. Comparable facility size, climate, set of unit processes and National Pollutant Discharge Elimination System requirements will be important. Corrections can be made to adjust for differing factors and unitize parametric values (e.g., per flow, per influent load, per total dry solids, etc.) for ease of comparison; and
   - Prepare a comparable estimate for your facility using the parametric information.
2. What is the industry "best practice" level of performance for a similar facility or a similar unit process? (*Repeat step 1 to see how industry sustainability leaders are doing.*)
3. How much of a given commodity (e.g., electrical power, heating/cooling energy, potable water, chemicals, etc.) are we consuming?
   - Gather the most relevant data (e.g., last 3 years); average by month or season to evaluated seasonal patterns; this is your benchmark or starting point;
   - Compare your results to the estimates from 1 and 2 above; and
   - Compare your results to internal standards, adopted levels of service, and sustainability policies.
4. Where are gaps in performance or opportunities for improvement? Use this information to prioritize projects and to set project goals that are realistic and measurable.

Benchmarking and gap analysis assist in identifying the utility's strengths and weaknesses as it relates to sustainability objectives. Figure 3 offers a high-level estimate of facility energy consumption based on facility type and average flowrate. Table 3 provides a list of energy consumption by unit process for industry average and "best practice" leaders. Recommended benchmarking resources include

- ENERGY STAR Portfolio Manager for Wastewater Treatment Facilities (U.S. EPA, 2016);

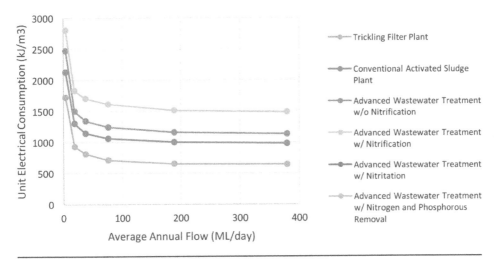

FIGURE 3   Benchmarking by facility type and flow.

- Electricity Use and Management in the Municipal Water Supply and Wastewater Industries (EPRI, 2013);
- Energy Conservation in Water and Wastewater Facilities (MOP 32; WEF, 2010);
- Utilities of the Future Energy Findings (ENER6C13) (WERF, 2014);
- Database of Best Practices for Energy Efficiency (compilation of case studies) (Water Research Foundation, 2012);
- Library of Water and Wastewater Technical Reports (NYSERDA, 2016); and
- Water and Wastewater Energy Efficiency (CEC, 2011).

3.1.1.1.3   ENERGY AUDITS

Energy audits can be performed to gather benchmarking information about the facility and to identify a range of projects to conserve energy. The main goals of an energy audit are to

- Identify energy use optimization improvements related to facility processes;
- Evaluate facility processes for potential energy use reduction and cost savings;
- Develop preliminary construction cost estimates;
- Develop estimated project cost savings items, including material/service contracts costs and energy costs; and
- Summarize simple payback and net present value economic analysis.

TABLE 3   On-site energy production unit estimates.

| Unit processes | Unit electrical energy consumption (kWh/ML)* | | Comment | Reference |
	Typical	Best practice		
Liquid stream treatment				
Influent pumping	55.6	39	210 kWh/MG (typical) and 148 kWh/MG (best)	(WE&RF, 2015)
Primary treatment				
Grit removal	7.4	5.8	Aerated grit (typical), forced vortex (best practice) (average consumption at 37 854 m³/d (10-mgd) facility)	(EPRI, 2013)
Primary clarifiers	7.9	6.7	Typical based on average consumption at 37 854 m³/d (10-mgd) facility. Best practice estimated.	(EPRI, 2013) Estimated
Ballasted sedimentation	20	16.9	Typical based on average consumption at 37 854 m³/d (10-mgd) facility. Best practice estimated.	(EPRI, 2013) Estimated
Secondary treatment				
Trickling filter/ solids contact	134.1		Typical based on average consumption at 37 854 m³/d (10-mgd) facility. No 'best-practice' for this process.	(EPRI, 2013)
Activated sludge: bio-chemical oxygen demand (BOD) removal only	190.5	93.3	Typical based on average consumption at 37 854 to 946 353 m³/d (10- to 250-mgd) facility. Best practice based on 51% improvement.	(EPRI, 2013) (WE&RF, 2015)
Activated sludge with nitrification (BNR)	285.7	142.9	Typical based on average consumption at 37 854 to 946 353 m³/d (10- to 250-mgd) facility. Best practice based on 50% improvement.	(EPRI, 2013) (WE&RF, 2015)
Activated sludge biological nutrient removal (BNR) mixing	29.1		Typical based on average consumption at 37 854 to 946 353 m³/d (10- to 250-mgd) facility.	(EPRI, 2013)

(*continued*)

TABLE 3   On-site energy production unit estimates (*Continued*).

Unit processes	Unit electrical energy consumption (kWh/ML)*		Comment	Reference
	Typical	**Best practice**		
Secondary treatment (*continued*)				
Membrane bioreactor	715.9		Typical based on average consumption at 37 854 m³/d (10-mgd) facility.	(EPRI, 2013)
Sequencing batch reactors	288.4		Typical based on average consumption at 37 854 m³/d (10-mgd) facility.	(EPRI, 2013)
Secondary clarifier	18.5	13.505	Typical based on average consumption at 37 854 to 946 353 m³/d (10- to 250-mgd) facility. Best practice based on 27% improvement.	(EPRI, 2013) (WE&RF, 2015)
Sidestream treatment				
De-ammonification (N-removal)		2.4	89 kWh/d at 37 854 m³/d (10-mgd) facility.	(WE&RF, 2015)
Struvite recovery (P-recovery)	8.5	6.7	250–320 kWh/d per Ostara Pearl 500 reactor. Pearl 500 can service 37 854 m³/d (10-mgd) facility.	(Seymour, 2009)
Effluent disinfection				
UV disinfection	67.5	16	Typical value based on UV-MP at 68 137 m³/d (18 mgd) (4560 kWh/d), Best-Case UV-LPHO at 68 137 m³/d (18 mgd) facility (1080 kWh/d).	(WEF, 2010)
Ozonation	66.7	26.7	Range of 100 to 400 kWh/MG. Typical value taken from middle, best-case from lowest.	(WEF, 2010)
Chlorine (on-site sodium hypochlorite solution generation)	28.1	11.22	Range: 42.5 (best case) to 170 kWh/MG.	(WEF, 2010)

(continued)

TABLE 3   On-site energy production unit estimates (*Continued*).

Unit processes	Unit electrical energy consumption (kWh/ML)*		Comment	Reference
	Typical	Best practice		
Effluent disinfection (*continued*)				
Tertiary filtration	15.3	7.7	Typical is for depth filtration at 37 854 to 946 353 m³/d (10- to 250-mgd) facility, 'best practice' is surface filtration at 37 854 to 946 353 m³/d (10- to 250-mgd) facility.	(EPRI, 2013)
Solids treatment				
Sludge thickening				
Gravity thickening	6.3		Typical based on average consumption at 37 854 to 946 353 m³/d (10- to 250-mgd) facility. No 'best-practice' for this process.	(EPRI, 2013)
Centrifuge thickening	10.3	4.1	Typical based on average consumption at 37 854 m³/d (10-mgd) facility. Best practice estimated.	(EPRI, 2013) Estimated
Digestion				
Conventional anaerobic mesophilic	20	2.9	Range is 2.9–32 kWh/ML (11–122 kWh/mgd); average is 19.8 kWh/ML (75 kWh/mgd) facility flow.	(WE&RF, 2015)
Aerobic	264.6	100.5	Typical based on average consumption at 37 854 to 946 353 m³/d (10- to 250-mgd) facility. Best practice based on 38% energy reduction modeled in WERF publication (2015).	(EPRI, 2013) (WE&RF, 2015)
Sludge dewatering				
Centrifuge	69.6	27.8	Typical based on average consumption at 37 854 m³/d (10-mgd) facility. Best practice estimated.	(EPRI, 2013) Estimated

(*continued*)

TABLE 3   On-site energy production unit estimates (*Continued*).

Unit processes	Unit electrical energy consumption (kWh/ML)*		Comment	Reference
	Typical	Best practice		
Sludge dewatering (*continued*)				
Screw press	4.2	1.7	Typical based on average consumption at 37 854 m³/d (10-mgd) facility. Best practice estimated.	(EPRI, 2013) Estimated
Belt press	12.2	4.9	Typical based on average consumption at 37 854 m³/d (10-mgd) facility. Best practice estimated.	(EPRI, 2013) Estimated
Sludge drying				
Direct thermal dryer	46.7	33.3	175 kWh/MG (typical) 125 kWh/MG (best practice).	(WE&RF, 2015)
Indirect thermal dryer	25.0	17.8	Indirect dryers require 0.051 kW/kg-H_2O evaporated (0.023 kW/lb-H_2O), versus 0.095 kW/kg-H_2O evaporated (0.043 kW/lb-H_2O) for direct dryers.	(WE&RF, 2015)
Solar drying		11.7	44 kWh/MG (best practice).	(WE&RF, 2015)
Process air handling and odor control				
Average level	37		300 kWh/MG.	(WE&RF, 2015)
High level	79		318 kWh/ML.	(King County, 2016)
Extreme	158			Estimated
Building mechanical				
Facility lighting	16		600 kWh/d for lighting, based on a 37 854 m³/d (10-mgd) facility.	(WE&RF, 2015)
Facility heating/ cooling	79		5000 MJ/d for heating/ cooling, based on a 37 854 m³/d (10-mgd) facility.	(WE&RF, 2015)

*Table values reported in kWh/ML.
3.78 (kWh/ML) = 1 (kWh/MG).
3.6 (MJ/ML) = 1 (kWh/ML).

During the energy audit, the following information typically is required:

- Historical data from the facility Supervisory Control and Data Acquisition/control system, if available;
- Average flow and load data;
- Pump curves;
- Blower curves;
- Actual operational data (motor amperages);
- Electric/gas utility rates and bills;
- Diffuser manufacturer data; and
- Miscellaneous plan drawings from previous construction projects (piping configurations, aeration diffuser layouts, etc.).

Energy audits may be done with in-house staff, consulting firms, or via partnerships with local utilities or state and federal government offices. The quality and depth of an energy audit can vary widely depending on the experience of the auditor, so it is important to use an auditor with experience specific to the water sector. The State of Maine Department of Environmental Protection prepared a sample RFP to help water sector facilities identify qualified energy auditors (Maine DEP, 2010). In Colorado, the governor's energy office offers energy audits for local governments and provides a list of pre-approved companies that conduct energy performance contracting and execute capital projects aimed at reducing energy consumption.

3.1.1.1.4 ECO-CHARRETTES

Eco-charrettes—or sustainability brainstorming workshops—are a component of integrated design (ID) and are intended to be collaborative creative sessions that seek to understand options for improving overall sustainability of the project and the constraints perceived by stakeholder groups. These workshops include brainstorming to establish sustainability development goals, strategies, and integrated design solutions. Conducting an eco-charrette early in the design process allows all team members and even outside stakeholders to contribute ideas. This activity increases engagement and generally improves buy-in by establishing a unified vision. Eco-charrettes are useful to uncover the utility's opportunities to improve sustainability and identify threats (risks) to continuing with status quo approaches. Group discussion and public input within an eco-charrette may result in identification of existing regulations, perceived constraints, or current policies that appear to conflict with the utility's or the community's sustainability goals.

Eco-charrettes require sufficient planning, including the use of a facilitator on larger or more complex projects, to get the best results. Conducting the eco-charrette in an offsite location to minimize typical workday distractions promotes participant engagement. The duration of the eco-charrette meeting or workshop depends on project size and complexity, and can be from 1 to 2 hours for small projects to 2 days with several breakout sessions for large, complex projects. Representatives involved in the design, construction, and operation of the project should participate, including operators, managers, engineers, code compliance, and other relevant parties.

3.1.1.2 Sample Set of Sustainability-Oriented Goals and Objectives

Sustainability goals should be realistic and objectives should be measurable. Data from benchmarking can be used to establish specific numerical goals. Example sustainability goals and objectives established for the installed project as compared to the baseline values for the current WRRF may include the following:

- Improve energy efficiency
  - Reduce purchased electricity per unit flow treated by ≥ 25%,
  - Reduce purchased natural gas per unit dry solids (loading) by ≥ 10%, and
  - Meet 25% of WRRF's needs via on-site production.
- Reduce GHG emissions
  - Reduce GHG emission per unit flow treated by 25%,
  - Reduce GHG emissions associated with transportation to/from site by 35%, and
  - Beneficially use ≥ 90% of biogas generated on-site.
- Conserve potable water
  - Reduce quantity of potable water used on-site by ≥ 50%,
  - Increase production of reclaimed water by ≥ 15%, and
  - Increase beneficial use of reclaimed water by ≥ 25%.
- Improve community understanding and endorsement
  - Increase participation levels in wastewater project outreach efforts by ≥ 10% (e.g., above current average of 10 persons per meeting or 4 comments per environmental review);
  - Identify relevant regulations or policies that appear to conflict with the community's sustainability goals for the project and suggest options to obtain variance or initiate change; and

- Seek industry partnerships and new revenue contracts by communicating resource recovery capabilities to local industries.
- Technical project goals (e.g., replace aged thickening and dewatering systems)
  - Provide sufficient system capacity under projected maximum month conditions through a 30-year service life;
  - Improve treatment performance reliability;
  - Select robust technologies with ease of O&M (e.g., minimize labor hours, frequency of servicing, availability of replacement parts);
  - Minimize requirements for purchased chemicals;
  - Include provisions for technology migration pathways; and
  - Improve resilience to short-term and long-term threats (e.g., power outages, seismic events, weather/climate change).
- Financial project goals
  - Minimize capital outlay for utility/owner to ≤ budget (total project, include construction and soft costs) and
  - Minimize operations and maintenance cost over 30-year period.

Technical goals and objectives generally relate the key project driver (e.g., asset replacement, capacity demand or growth, regulatory requirements, and market drivers). Site-specific opportunities (e.g., to recover and re-use or sell resources) or threats (e.g., industry relocation) may also be drivers that influence project goals. During goal setting, it is useful to also recognize the relative risk tolerance of the utility and/or include a "risk budget" when considering new technologies or innovative solutions. It is essential that there is widespread buy-in for the goals and objectives. Without team member buy-in and support from operations and/or upper management, success is unlikely.

3.1.2 Siting Decisions

Within a given service area, the specific location of a WRRF has substantial effect on the resulting overall efficiency of operation and long-term sustainability. Considerations and recommendations for siting decisions (new facilities) include near-shore versus upland, site size, number of facilities, orientation on site, hydraulic profile, community input, and stormwater management.

3.1.3 Community Integration

Community integration is a key element to long-term sustainability. The following list provides some examples of how wastewater agencies can integrate sustainability principles and public amenities into WRRFs in ways that increase the value of having a facility in the neighborhood:

- High-tech training centers that can be used for staff training and also rented out to offer skills-building training convenient to the community;
- Science/environmental learning centers to raise awareness of environmental stewardship, reduce water consumption, and change water polluting behaviors;
- Tour routes, signage, and other educational opportunities for school tours;
- Art enhancements for new construction that engage local artists (e.g., integrated artwork, temporary artwork along construction fences);
- Maintenance shop with training stations for plumbing and building mechanical (e.g., pressure reducing valves, backflow preventers, pumps, chillers, etc.) to support vocational training;
- Hands-on laboratory that offers classes for high school and college level science, technology, engineering, and mathematics students to inspire next generations into water and wastewater careers;
- Parks and playfields using reclaimed water and recovered fertilizer or biosolids;
- Transportation connections through/along facility property to improve neighborhood connectivity and access to recreation;
- Hot water energy districts that recover "waste heat" and offer it at competitive prices to nearby businesses to encourage commercial development;
- Reclaimed water to nearby industry for purple pipe uses such as landscape irrigation;
- Export surplus electricity at competitive "green" prices to reduce community dependence on non-renewable sources; and
- Export surplus biomethane as renewable natural gas or compressed natural gas (CNG) to reduce community dependence on non-renewable sources.

3.1.4 Opportunity Components

In the course of scoping the project, the team should ask: "What additional improvements might be included at zero/low marginal cost that would improve sustainability of the resulting installation?" Examples of opportunity components may include

- Repair/replacement work—Replacement of motors in existing facilities with the installation of premium efficiency motors may be justifiable using a life cycle cost assessment. Work to repair leaking air and facility water lines may be an easy add-in to save potable water and energy;

- Lighting—Maximize use of natural light by including a skylight or via smart window placement; use light-emitting diode (LED) lights for lower energy consumption; power LED lights via photovoltaic panels. For site lighting, determine early if full cut-off exterior lighting is desired or required by local code (e.g., to reduce light pollution and support dark-sky goals) and plan ahead for a greater number of lights;

- Water reuse—Maximize opportunities for production and use of reclaimed water (e.g., plumb toilets, process uses, sale to nearby industry); discuss partnership opportunities with local industry early;

- Heating systems—Consider all options to improve heating system energy efficiency for process loops and buildings. Make provisions to loop and interconnect hot water heating systems and include heat recovery on sludge processes. Consider effluent heat recovery or other approaches to beneficially capture and use waste heat energy;

- Greenhouse gas emissions—Evaluate options to cover tanks, and improve air handling, odor control, and emissions;

- Biological habitat—Consider options to enhance compatible habitat on-site (e.g., honey bees, butterflies, etc.) and that preserve wildlife corridors through or adjacent to the site; and

- Public amenity—See the previous section for examples of community integration components.

3.2   Preliminary Design Phase

This phase encompasses conceptual design (10 to 15% level) and includes use of an integrated design approach, selection of feasible technologies, alternatives screening, evaluation, and selection. Activities and factors that influence selection of treatment technologies that will improve energy management and overall sustainability are described in this section.

3.2.1   Integrated Design for Sustainability

Integrated design is rooted in sustainability because it is highly collaborative and requires the project team to think of the entire facility or process, its O&M, and its interfaces with and potential effects on adjoining processes and/or surroundings. Integrated design recognizes that facilities or processes are part of larger, complex systems or surroundings. If LEED or Envision awards will be pursued, ID documentation efforts and eco-charrettes begin during the planning phase and will continue through design and construction phase. The use of the ID process affirms the team's commitment to sustainability.

3.2.2 Technology Selections

Process selection can dramatically affect energy demand requirements. Several of the more energy-intensive unit processes within a WRRF include

- Wastewater pumping;
- Secondary treatment aeration;
- Thermal hydrolysis;
- Digester heating;
- Ultraviolet (UV) disinfection;
- Dissolved air flotation thickening;
- Centrifuge thickening or dewatering;
- Building heating, ventilation, and air conditioning; and
- Odor control.

Technologies that reduce energy consumption for these unit processes will have the biggest effect on a WRRF's overall energy balance. Recommendations for technology selections to conserve energy and to recover/produce more energy on-site are highlighted here. Refer to *Design of Water Resource Recovery Facilities* (WEF et al., 2018) for further details about specific unit processes.

3.2.2.1 Energy Conserving Technologies

For new facilities or projects considering adding new processes or for facilities considering energy-focused retrofits, the following technologies to reduce energy consumption are suggested for consideration:

- *Liquid Stream*
  - Pumping—Centrifugal pump manufacturers offer innovative impeller designs that reduce energy consumption. Optimizing pumping throughout the facility has been reported to reduce energy consumption by an average of 0.7% (WERF, 2011).
  - Advanced primary treatment—A variety of methods, including chemically enhanced primary treatment, dissolved air floatation,

microscreens, and proprietary filters are being used to divert carbon from the influent to the solids process. This has the double benefit of reducing downstream aeration energy demands and enhancing energy recovery from solids processes like anaerobic digestion (Fitzpatrick et al., 2015).

○ Optimization of conventional activated sludge for biological nutrient removal. A variety of optimizations have been implemented with good results. The following are examples:

- Aeration system optimization—Aeration system optimization (e.g., energy-efficient aeration systems, improved control over operating levels of dissolved oxygen) has been reported to reduce energy consumption by 15 to 38% (WERF, 2011).

- Flexible zoning—Addition of flexible sequencing of basins (e.g., swing zones) and/or with seasonable adjustments to operation strategy has been reported to reduce energy consumption by 8 to 22% (WERF, 2011).

- Anaerobic selectors—Addition of an anaerobic zone to the first cell of activated sludge selects for phosphorus-accumulating organisms, reduce aeration demands, improve solids settleability, and increase phosphorus content of waste activated sludge (WAS). This approach pairs well with sidestream phosphorus treatment.

- Modified Ludzack–Ettinger (MLE) Process—At facilities that require nitrification and denitrification, the MLE process recovers the oxygen value from nitrate and can reduce total air demand by 15 to 30%.

- Biomass retention technologies—System configurations that retain biomass and/or de-couple hydraulic residence time from sludge residence time (SRT) have several energy advantages. This approach allows for higher loadings rates per unit volume and may be a sustainable change. Examples include sequencing batch reactor, moving bed biofilm reactor, integrated fixed film activated sludge, and Bio-Mag.

○ High-efficiency ultraviolet disinfection—Ultraviolet disinfection is an excellent option for disinfection of wastewater, but conventional UV systems typically require more power as compared to chlorination. Recent innovations in UV technology (e.g., low-intensity bulbs, electronic ballasts) and improved control over power output or pulsed technologies should be evaluated as design options when considering UV.

○ Reclamation facilities—Tertiary treatment to produce reclaimed water is an added process and thus will increase energy demand. The preferred process will vary depending on end-use expectations. Membrane bioreactors (MBRs) can produce a high-quality effluent but require considerably more mechanical equipment (e.g., more blowers, pumps, and backwash systems) and require more air due to the longer sludge age and nitrification requirement for optimal membrane flux. As a result, MBR systems have significantly higher energy consumption and greater maintenance requirements compared to conventional activated sludge systems. In addition, MBR systems may actually improve the effluent quality more than is necessary for the intended use. Filtration processes (e.g., disk filters) will typically have lower electrical energy demand and result in reclaimed water suitable in quality for most Class A uses.

- *Solids Stream*

  ○ Solids pretreatment—Technologies that significantly improve thickening and/or increase cell lysing to allow for greater loading rates to anaerobic digestion will typically increase biogas production and offer greater potential for net on-site energy recovery. There are a variety of unit processes and range of net energy demands for this approach. Process modifications, intended to reduce the quantity of biological solids requiring disposal, should be assessed for the net energy that will be required (i.e., include pumping and aeration).

  ○ Upflow anaerobic sludge blanket reactors (UASBs)—UASBs are effective for treating higher strength influent streams and can produce a considerable amount of biogas in a relatively small footprint. This process generally performs best when operated year-round. Upflow anaerobic sludge blanket reactors are commonly used in the food and beverage industry in the United States and throughout Europe.

  ○ Biosolids dryers—Mechanical sludge dryers that increase the solids content to 90% or more produce excellent residual biosolids products and reduce energy for hauling; however, evaporating water in this manner is very energy intensive. Some facilities, including the Greater Lawrence Sanitary District WRRF in Andover, Massachusetts, use biogas as an energy source to offset natural gas demand for drying solids. Market location for biosolids reuse is a key consideration. A comparison of cost and energy saved should be made to determine if mechanical biosolids dryers are appropriate and to select the best drying technology.

- *Sidestream*
  - Nitrogen treatment—Sidestream treatment processes are useful to reduce the amount of ammonia returned to aeration basins solids processing and thereby reduce aeration requirements and effluent ammonia concentrations. There are several new technologies for sidestream treatment (e.g., DEMON, anammox). Current practice is to treat concentrated dewatering liquor in a sidestream process before blending it with the primary effluent.
  - Phosphorus treatment—Strategies that concentrate and remove phosphorus from sidestreams have several sustainability advantages. Waste activated sludge pre-treatment and dewatering of digestate releases soluble phosphorus (in the form of orthophosphate), which can be harvested in the form of struvite (magnesium ammonium phosphate). This also offers a positive means to control nuisance struvite deposits. A symbiotic process pairing would be to add an anaerobic cell to the first zone of activated sludge to select for phosphorus-accumulating organisms, reduce aeration demands, improve solids settleability, and increase phosphorus content of WAS. This example paired approach reduces energy consumption, produces a valuable fertilizer byproduct, and improves reliability of operations.
- *Facility-Wide*
  - Lighting improvements—Improvements to use more natural light and/or replace incandescent lighting with LED lighting in buildings and site lighting has been reported to reduce energy consumption by 2 to 6% (WERF, 2011).
  - Heating, ventilation, and air conditioning—In warm climates, thoughtful building and window orientation, variable refrigerant flow technologies and alternatives to electric powered air conditioning should be considered. In cold climates, space heating can be a big component of facility energy consumption. Several strategies to reduce heating demand include
    - Air-to-air heat exchangers (e.g., pre-warm incoming air with exhaust air);
    - Influent or effluent heat pumps that use electricity to transfer heat from a cold fluid (e.g., effluent) and deliver it to a warmer fluid (e.g., closed hot water loop); and
    - Solar thermal heating for south-facing spaces can also be used. The Saco WRRF also uses two types of solar thermal heating and heat pumps (Day, 2013).

- *Innovative and Emerging Technologies*
  - Innovative low energy secondary treatment—System configurations that support growth of anammox bacteria, retain nitrifiers, and/or otherwise reduce aeration demands are promising. Examples include the A-B process or granular sludge.
  - Algae bioreactors—Algae naturally grows in nutrient-rich streams including wastewater and can be further processed into biofuels (e.g., biodiesel, ethanol). Research into algae species selection and cultivation technologies to reduce footprint and increase yield is showing promise.
  - Microbial fuel cells—Microbial fuel cells generate electricity and hydrogen peroxide from organics present in wastewater. Application of this technology at full-scale is not yet practiced, although researchers continue to make advances that may lead to marketable products in the near future.

3.2.2.2 Energy Producing Technologies

Technologies that produce heat and/or power include the following:

- Anaerobic digestion—Anaerobic digestion converts chemical oxygen demand (COD) in the wastewater into methane and carbon dioxide; methane can be captured and used to satisfy on-site demands for heat and power.
- Co-digestion—Form of expanded anaerobic digestion, which includes additional co-processing of supplemental organic substrates received from outside sources (e.g., fats, oils, and greases [FOG], municipal food waste, industrial food and beverage waste, etc.). Accepting such feedstocks can increase production of biogas, improve the overall facility energy balance, and bring in tipping fee revenues (U.S. EPA, 2014).
- Biogas utilization processes—Combined heat and power, biogas-fired boilers, biogas-driven engines powering pumps and blowers, biomethane recovery for pipeline injection, and CNG for fleet vehicles are examples of ways to beneficially use biogas and minimize flaring.
- Electricity from thermal processing—Facilities with incineration may be able to add electricity generation from the waste heat at the incinerator using a variety of generator designs. Reference facilities include the Metropolitan District Hartford, Connecticut, and one in Albany, New York.

- Alternative energy components—Photovoltaic and solar heating components may be suitable for installation on roof and facility areas with sufficient solar gain. Wind turbines may be appropriate if wind patterns are suitable. For facilities with considerable hydraulic drop, it may also be possible to install a hydroelectric generator on the effluent outfall.

3.2.2.3 Key Factors in Selecting Technologies and Preferred Alternatives

Key factors in selecting best-fit technologies include effluent permit requirements, location and climate, and relative facility size. In some cases, the potential for water quality trading may be important.

- Effluent requirements—Effluent requirements of the WRRF have a direct effect on the potential to achieve specific energy and sustainability goals and objectives. Conventional technologies generally trend with increasing energy use and increasing GHG emissions as facilities move toward greater levels of nutrient removal (nitrogen, phosphorus). For nitrification facilities, there are sustainability benefits of running with a longer SRT and including denitrification in the process train. Note, however, there are significant tradeoffs in terms of sustainability for the higher levels of nutrient removal using currently available technology. A 2011 study by the Water Environment Research Foundation (WERF) found a point of diminishing returns may be reached as effluent limits approach Level 4 (3 mg nitrogen/L; 0.1 mg phosphorus/L effluent target) resulting from the added cost, energy use, chemical use, and GHG emissions (70%) with only a relatively small reduction in nutrients (1%) (WERF, 2011). The results of that study suggest that Level 3 or 4 treatments complimented with best management practices of non-point sources is a more sustainable approach at achieving comparable water quality, as compared to requiring point source dischargers to achieve Level 4 or 5 treatments (WERF, 2011).

- Location and climate—The location and climate of each WRRF directly affects the influent temperature, liquid stream process kinetic rates, building air heating and cooling demands, feasibility for alternative energy production, and emissions and odor control requirements. The location affects the availability and unit cost for natural gas or propane fuel, proximity to a gas pipeline, and driving distance relative to local industrial dischargers or resource recovery partners. It also dictates the cost for electrical power and the GHG footprint of this power supply. Technology options should be short-listed based on applicable geographic factors. The design team should evaluate the WRRF's seasonal energy balance and look for technologies that would balance energy demands with potential to produce on-site. This entails comparing monthly and possibly diurnal swings in flows, loads, and operational strategy; understanding utilities pricing across seasonal and daily high–low demand periods; looking for symbiotic pairings of technologies; and considering site layout, building orientation, and hydraulics.

- Facility size—There are economies of scale offered to facilities both large and small. Consider the capacity of commercially available equipment for a given technology. For small facilities, the feasibility for anaerobic digestion may be increased if FOG receiving is included or if there are available organics from nearby industrial dischargers.

- Potential for water quality trading—In areas with total maximum daily limits or other regulatory requirements that support water quality trading, opportunities to use off-site strategies to cut pollutant loadings (e.g., phosphorus, metals, emerging contaminants) to receiving waters should be evaluated. In some cases, these strategies may be more cost-effective than designing the facility to meeting higher effluent quality limits. Designers should weigh the pros and cons of both approaches and strive to optimize social, economic, and environmental benefits. Interest in the trading approach is increasing, as communities seek more holistic solutions. Water resource recovery facility managers can be leaders in this regard—by educating customers and offering a financial mechanism via the rate structure and bonding capacity—to help local communities meet their pollution reduction goals (Stoner, 2011).

3.3   Final Design Phase

During the final design phase, decisions are made about how the various forms of energy entering the facility are made, as well as selection of specific equipment, system layout, hydraulic efficiency, and provisions for future technologies. The result of design decisions fix the ultimate boundaries of energy efficiency and long-term sustainability of operations for the resulting facility. Thus, it is vital to build sustainability early into the design process and not attempt to apply as an afterthought. Energy modeling, GHG modeling, and life cycle costing are described in this section. The ID process continues through this phase and requires ongoing engagement of the owner, designer, architect, regulators, and community stakeholders. Design milestones should include a review of progress toward sustainability goals and objectives.

3.3.1 Energy Modeling

Energy modeling is a tool to assist the designer in assessing the net effect on energy intensity of several design options. Calculations are performed during predesign and final design to evaluate the facility-wide balance of energy sources and uses for design options. Modeling can be used to assess the energy consumption for a particular piece of equipment, system, process, or facility. The estimating method and tools will depend on factors such as the desired output (cost or units of energy), purpose or value of the information being sought, degree of accuracy needed, how quickly the estimate is needed, potential effect on cost or savings by knowing the estimate, and amount of effort or budget available for conducting the modeling.

Energy modeling tracks the flows of energy in and out of a WRRF. The goal is to balance the energy demand with the energy supply available on-site. *Zero net energy* refers to a WRRF's ability to create on-site all of the electrical and heating fuel energy needed to sustain operations. A zero net energy facility may purchase power and natural gas during colder months while exporting energy during other times of the year; however, on average the annual quantity of energy produced on-site is about equal to the total annual average quantity used by facility operations. A preliminary estimate of the facility-wide energy balance can be prepared at approximately 30% design and finalized by 90% design. *Sankey diagrams* are often used for this; these are graphical depictions of how various forms of energy (chemical/carbon, electricity, natural gas/biogas, and thermal) flow into the WRRF and through the major unit processes. There are a many software tools available that can be used to plot a Sankey graphic.

Energy is tracked in several forms. The energy content of influent wastewater includes:

- Thermal energy—Heat energy contained in the wastewater, which is governed by the specific heat capacity of water (e.g., $41\ 900\ \text{MJ}/10^3 \cdot \text{m}^3$ for every 10 °C);
- Hydraulic energy—Hydraulic energy comes in two forms:
  - Potential energy is the energy resulting from the water elevation at the tine inlet to the WRRF relative to the outfall disposal elevation. This "head" is a motive force that should be used efficiently to drive the water through the liquid stream treatment processes and to effluent disposal. Some facilities may be able to harness excess potential energy via a generator on the outfall pipe; and
  - Kinetic energy is the energy from moving water (velocity).
- Chemical energy is the energy content stored in the various organic chemicals in the wastewater. This energy is typically calculated on

COD or on solids loading (e.g., 13 MJ/kg COD; 20 MJ/kg dry solids in influent total suspended solids). Chemical energy drops through the liquid stream treatment process, as carbon is converted into carbon dioxide (which offgases) and also transferred to the solids stream in the sludge.

The energy used at WRRFs is generally in the form of the following:

- Electricity—either purchased form the local utility or produced on site. Electrical demand can be reported in terms of kilowatt hours per day (kWh/d) or megajoules per day (MJ/d);
- Heating fuel—this can be electricity, natural gas, propane, or fuel oil. Also, heat pumps driven by electricity can be used to deliver heat from treated effluent. For facilities with anaerobic digestion, heating fuel would include digester gas, which is, on average, approximately 65% methane. Heat can also be transferred for air, water, and sludge streams; and
- Gas-powered engines—the motive force to move water or pressurize air may also be derived from biogas or natural gas. The biogas-fueled raw wastewater engine pumps at the West Point Treatment Plant operated by King County (Seattle, Washington) are an example.

Sankey diagrams are graphical tools to explain how the various forms of energy flow through a WRRF. Figures 4 and 5 are Sankey diagrams that represent energy flows into, within, and out of a model activated sludge WRRF with secondary treatment. The first figure shows a typical facility with anaerobic digestion and CHP. The second figure depicts the same facility with maximal energy efficiency, energy recovery, and the digester augmented with outside energy sources (fats, oils, and grease; food waste). Energy modeling can be done via hand calculations or via use of commercially available software. Example models for WRRFs include BIOWIN V4.1 (EnviroSim, 2016), ENERC1 (WE&RF, 2015), and ENERGY STAR (U.S. EPA, 2016). Commercially available Sankey graphing software is also available online, including open source options.

3.3.2 Greenhouse Gas Modeling

Environmental cost accounting is a growing field that includes methods to estimate and value GHG emissions. Accounting for GHGs from a WRRF is typically done as a two-step process. Step 1 estimates direct emissions from facility operations. Step 2 estimates indirect emissions from related

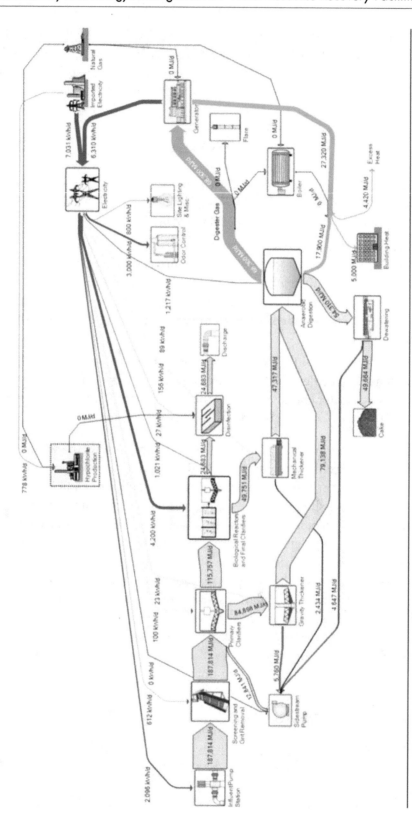

FIGURE 4   Sankey diagram 1 (WE&RF, 2015).

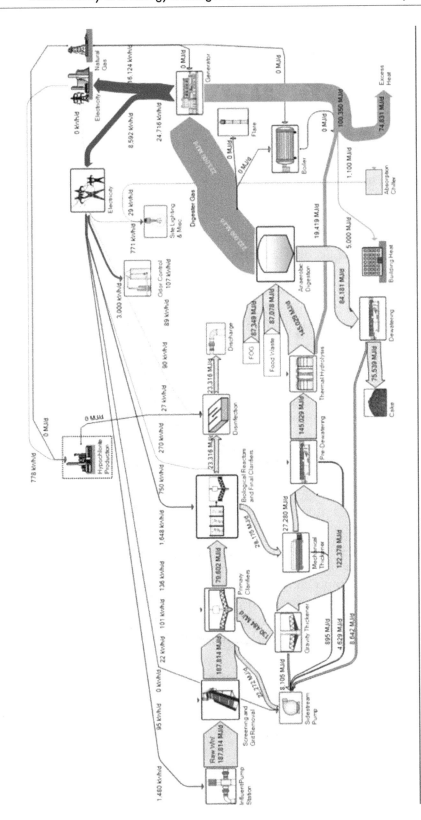

FIGURE 5   Sankey diagram 2 (WE&RF, 2015).

sources outside the facility boundary. The primary contributors to step 1 GHG emissions are energy related and include high electric uses at the facility where local power is fossil fuel dominant. Such processes include aeration, pumping, and mixing. Primary contributors also include emissions associated with high heating demands for both process and space heating, as well as process-related emissions that include methane released from the collection system and in biogas flaring and CHP facilities, as well as volatile organic compounds and nitrous oxides from secondary treatment aeration tanks. Step 2 emissions would include chemical import and use and biosolids hauling. Green house gas emissions for processes requiring chemicals are typically higher than those that do not use chemicals. Export of recovered resources would be accounted for under step 2.

Greenhouse gas accounting protocols are useful guides for developing GHG emissions inventories, but engineers should not be overly reliant on the protocols as they are evolving and improving rapidly. For a rough approximation, the maximum CO_2 generation rate based on facility loading can be approximated as 44 kg CO_2 per 32 kg five-day BOD (WEF, 2010). Additional level of analysis requires more rigorous evaluation of unit processes and utility power and natural gas consumption.

Willis et al. (2016) describe several protocols currently in use to assess direct (step 1) GHG's emissions from WRRFs. These protocols have addressed emissions from wastewater conveyance and treatment using a variety of simplifying methodologies. This review summarizes the sources of direct GHG emissions (both those covered and not covered in current protocols) from wastewater handling; provides a review of the wastewater-related methodologies in a select group of popular protocols; and discusses where research has outpaced protocol methodologies and other areas in which the supporting science is relatively weak and warrants further exploration. These are summarized in Table 4.

Because GHG emissions per unit of electric vary substantially depending on the power supply mix (i.e., coal, hydroelectric, etc.), the most accurate approach will be to contact the local electrical power provider for the project site to determine the carbon footprint of the local power mix (i.e., tons CO_2eq/kW·h). Resources for estimating GHGs from WRRFs include:

- ENERGY STAR Portfolio Manager for Wastewater Treatment Plants (U.S. EPA, 2016) can calculate greenhouse gas emissions for electricity, natural gas, and fuel oil consumption at a facility, adjusted for the assumed local electric fuel mix based on a facility ZIP code. Because this tool uses emissions factors based on facility ZIP codes, it will be the most accurate but it requires more data input than the following options. See https://www.energystar.gov;

TABLE 4   GHG modeling protocols by treatment process (Willis et al., 2016).*

Treatment process	Unit processes	Primary GHG emissions	Available GHG protocols
Influent pumping	Collection system sewers, pumping	CH_4	IPCC
Preliminary treatment	Headworks processes (screening, grit removal)	CH_4	NGA/NGER
Primary treatment	Primary clarification	CH_4	NGA/NGER
Secondary treatment			
Aerobic secondary treatment	Biological treatment reactors, solids treatment	CO_2 (from COD reduction)	IPCC
Aerobic treatment for nutrient removal	Denitrification treatment reactors	CO_2 (from methanol consumption)	IPCC, ICLEI's U.S. Protocol
Sidestream/ mainstream nitrogen removal (aerobic)	Biological treatment reactors	N_2O	IPCC, CARB's LGOP, ICLEI's U.S. Protocol, NGA/NGER
Secondary clarification	Secondary clarifiers	N_2O	NGA/NGER
Effluent discharge to receiving water	Outfall	N_2O	IPCC, CARB's LGOP, ICLEI's U.S. Protocol, NGA/NGER
Thickening	Thickening equipment	CH_4, CO_2	—
Anaerobic digestion	Anaerobic digester	CH_4, CO_2	IPCC, NGA/NGER
Aerobic digestion	Aerobic digester	CO_2	IPCC, NGA/NGER
Dewatering	Dewatering equipment	CH_4, CO_2	—
Biogas combustion	CHP, boilers, waste gas burner	CH_4	IPCC, CARB's LGOP, ICLEI's U.S. Protocol, NGA/NGER
Solids disposition	Land application of digested biosolids, combustion of biosolids	CH_4, N_2O	IPCC, CARB's LGOP, ICLEI's U.S. Protocol, NGA/NGER

Notes: See Willis et al. (2016) for more information. Abbreviations above are as follows:
Intergovernmental Panel on Climate Change (IPCC) (2006); *Local Government Operations Protocol* (LGOP) by the California Air Resources Board (CARB) (2008); *U.S. Community Protocol*. International Council for Local Environmental Initiatives—Local Governments for Sustainability USA (ICLEI) (2012); and *Australian National Greenhouse Accounts Factors* (NGA) and *National Greenhouse and Energy Reporting System* (NGER) (2013).

- U.S. EPA's Greenhouse Gas Calculator: https://www.epa.gov/energy/ greenhouse-gas-equivalencies-calculator. This calculator converts tons CO_2eq to common terms such as the equivalent number of homes heated per year or gallons of gasoline. This method relies on national averages for power mix. Although it is less accurate resulting from its reliance on a national average, it is the easiest tool to use and is adequate for quick estimates; and

- Australian Government's Clean Energy Regulator Free GHG online calculator (Microsoft Excel) for municipal and industrial WRRFs, included here for international users: http://www.cleanenergyregulator .gov.au/NGER/Forms-and-resources/Calculators.

3.3.3 Life Cycle Costing

Life cycle costing is an important approach to assessing long-term economic viability. This section offers the following suggestions regarding how to value energy and other sustainability related parameters. The following are some suggestions:

- Timing—Before initiating evaluation of alternatives, establish all parameters that will be valued (e.g., GHGs, biofuels credits, etc.), how they will be valued, and assumed economic factors (unit costs for purchased utilities, discount rates, etc.) to avoid biasing results; and

- Value of thermal energy—Although commodity unit prices are typically easy to identify from the local utilities, it can be more difficult to set the economic value of using (or not) "waste heat". Thermal energy capture is often larger than the chemical energy fraction but this is "low grade" energy and thus difficult to leverage for useful work in the way that electricity can be readily used. Depending on location, thermal energy source could be tapped to offset purchased utilities for digestion, space heating, hot water heating, process water, and industrial uses. Project examples that involved connections between a WRRF and a district heat utility or sale to an industrial buyer include those conducted by King County, Washington, Avon, Colorado, and Vancouver, British Columbia.

3.3.4 Equipment Selection

3.3.4.1 Energy Management Information Systems

Water resource recovery facilities' managers are encouraged to include energy management information systems (EMIS) in design contracts. An

EMIS includes a combination software, data acquisition hardware, network communications, and instrumentation and automation that are used to monitor, control, and report energy consumption. An EMIS can be applied to a singular building or an entire industrial processing facility such as a WRRF. Capturing and displaying data in a graphical manner (e.g., energy dashboard) and producing reports that include energy consumption and production trends are all readily achievable using EMIS.

3.3.4.2 Pumping

The design of a water or wastewater pumping system will greatly affect energy use. A proper design should consider the energy implications of peak flowrates, pipe/force main sizes, and equipment selection. Careful attention should also be paid to the proper application of variable speed pumping and multi-stage pumping strategies.

Preliminary and final designs of pumping system projects should include details regarding the following energy efficiency considerations:

- Types of pumps that minimize energy use and maintenance problems, especially clogging with wastewater debris;
- Right size pumps, motors, and drives. Pump operating point selection and number of pumps needed to maximize energy efficiency for average operating conditions while meeting peak flow and reliability requirements;
- Including a small jockey pump may help some designs minimize energy consumption across wide flow ranges;
- Control strategies that function for the full range of operating conditions but are optimized for early years; plan to update set points as flows grow over time;
- Life cycle cost assessment of premium versus standard efficiency equipment, including energy and maintenance costs;
- Preventive maintenance schedules consistent with manufacturer's requirements;
- Require submittals of O&M manuals to state manufacturer's expected system efficiency and offer troubleshooting tips if actual performance drops below expectations;
- Require training to include energy management training of operations staff; and
- Monitoring that captures and reports data regarding system energy efficiency.

3.3.4.3   Aeration

Aeration in the secondary treatment process accounts for 30 to 60% of total energy consumption at the typical activated sludge WRRF (EPRI, 2013). There are a number of opportunities to improve the energy performance of aeration systems, including sensors and automated controls, fine bubble diffusers, efficient blowers, and variable speed drives. Guidance to designers and owners to improve energy efficiency of aeration systems is listed below:

- Designers—Evaluate latest technologies for diffusers, mixers, blowers and instrumentation, including consideration of
  - Predicted system performance across the wide range of operating conditions from startup to design year;
  - Wire to air efficiency (Standard CFM/kWh) of various types of blowers, including high-speed turbo blowers;
  - Appropriateness of multi-blower or cascading blower operational schemes;
  - Fine-pore or membrane-diffused air systems;
  - Use of high efficiency mixers and mixer-aerators;
  - Options for tighter dissolved oxygen control (e.g., optical sensors, redox probes); and
  - Functionality offered by "packaged" blower products (e.g., blower, variable speed drive, dissolved oxygen sensor, and programmable logic controllers from one supplier) versus custom-designed systems.
- Owners/Operators—Clearly define O&M procedures and standard operating practices (SOPs) that seek to maximize energy efficiency and system performance. These should include
  - Energy management training for O&M staff;
  - Equipment service schedules—including diffuser maintenance (cleaning, replacement of broken units, instrument calibration, etc.)—at frequencies recommended by manufacturers, or at least annually;
  - Monitoring and testing protocols; and
  - Annual testing to check blower and aeration system energy efficiency (e.g., amp draw, head loss, air flow, back pressure, etc.).

3.3.4.4   Solids Handling

The solids treatment train offers many opportunities to improve the energy balance of the facility and reduce the hauling and disposal costs of residual biosolids. Water resource recovery facilities are increasingly using anaerobic digestion of solids as a means to supply the onsite energy needs of the facility. In evaluating solids digestion, dewatering, and disposal options, include details regarding the following for each alternative:

- System energy performance calculations under a range of operating conditions (present conditions, growth scenarios);
- Life cycle cost assessment of premium versus standard efficiency equipment (including maintenance costs, extent of thickening or dewatering achieved, and resulting sludge hauling and disposal costs);
- Payback for variable speed drive technology;
- Options for sludge disposal or reuse other than current method;
- System monitoring and control;
- Equipment service schedule;
- For a facility with anaerobic digestion, options to move the facility closer to net zero for purchased energy include
  - High performance anaerobic digestion improvements that maximum throughput, biogas production, and volatile solids reduction;
  - Increasing methane production by importing and co-digesting additional organic substrates (e.g., FOG, food waste) from industry partners;
  - CHP production;
  - Recovery of biomethane for pipeline injection or CNG for fleet vehicles;
  - Steam-driven thermal hydrolysis paired with gas turbines;
  - Hot water driven thermal hydrolysis paired with internal combustion engines; and
  - Heat recovery via sludge cooling and sludge preheating hydronic heat exchangers.

3.3.4.5 *Ultraviolet Disinfection*

Ultraviolet light provides final effluent disinfection at many WRRFs. Although it is more energy intensive than disinfection, UV avoids the many issues associated with handling chlorine and managing unwanted chlorine byproducts of disinfection. Ultraviolet technology energy efficiency is improving, however, with features that include dose-pacing control, level-pacing control, and better system turndown.

3.3.5 Materials Selection

Sustainable material selection focuses on both minimizing the use of natural resources and the environmental effects of material production and use. Material durability and reliability, content and production, sustainable sourcing, and end of life considerations are important factors in materials selection. *Sustainable sourcing* means purchasing from suppliers who can document (e.g., via Environmental Product Declarations [EPDs] and/or Health Product Declarations [HPDs]) that their raw materials and resulting product was manufactured in a sustainable manner. Design teams ideally should weigh and compare life cycle effects to select the optimal material.

3.3.5.1 Durability and Reliability

Durability and reliability are core concepts of sustainable design that are intended to reduce waste resulting from equipment or component repair and replacement. *Design of Water Resource Recovery Facilities* (WEF et al., 2018) discusses the design considerations for WRRF environments, including corrosive, moist, and other potentially harsh conditions that impact material durability and reliability. Flexibility and ease of adaptation to changing conditions also maximizes equipment and component life.

3.3.5.2 Content and Production

Material content and production consider whether materials are produced from virgin or recycled stock and the environmental impact of the production process. Environmental Product Declarations (EPDs) and HPDs are used by the USGBC as part of its LEED rating systems to encourage use of materials that have fewer environmental, economic, equity, and human health effects. "The Red List" (International Living Future Institute, 2016) identifies chemicals and materials that are harmful to humans and should be avoided where possible. Thoughtful material selection can help minimize the generation of persistent bioaccumulative toxic chemicals by eliminating, reducing, or specifying substitutes for materials containing or whose manufacturing generates harmful substances such as mercury, lead, cadmium, and dioxins. Example material content and production considerations include

- Polyvinyl chloride (PVC)—PVCs and chlorinated PVCs are receiving increasing scrutiny because they are derived from non-renewable petrochemicals and the manufacturing process has been associated with release of dioxin (PPRC, 2015);
- High density polyethylene (HDPE)—HDPE and polyethylene piping can be used in place of PVC piping in certain applications. Although

HDPE may be petrochemical based, the HDPE manufacturing process is viewed by some as less detrimental than that of PVC (PPRC, 2015). It is also possible to obtain HDPE with recycled content;

- Volatile organic compounds (VOCs)—Low VOC emitting, low toxicity paints/finishes, sealants, caulks, adhesives, insulation, and furnishings/materials should be specified whenever possible;

- "Virgin materials"—Using virgin materials—those with no recycled content—results in higher embodied energy in the manufactured component and is a more environmentally disruptive selection than choosing materials with recycled content, or that are reclaimed, reused, or repurposed;

- Recycled materials—Using recycled materials, or materials with recycled content, reduces materials disposed of in landfills or incinerators. Metals, asphalt, aggregate, concrete, wood forms, architectural products, and some plastics are examples. Onsite use of excavated materials should be considered where feasible;

- Concrete with recycled content—Concrete with fly ash or furnace slag byproducts of combustion or production processes is readily available;

- Rapidly renewable materials—Bamboo, for example, in place of harvested lumber is considered rapidly renewable resulting from the fast growth rate of bamboo; and

- Buy local—Regionally sourced materials limit emissions resulting from journeys from far-away lands and can positively contribute to the local economy. Note, however, that rules for some federally funded projects in the United States may prohibit specifying regionally sourced materials, which may limit qualifying for some sustainability system credit.

Life cycle assessments (LCAs) are computer modeling tools for energy and GHGs that are used to evaluate material selection and embodied energy. Life cycle assessments are typically "cradle-to-grave" analyses of a product, assembly, or facility's environmental effect resulting from material extraction, manufacturing, transportation, installation, use, maintenance, and disposal. Life cycle assessments applied on materials used in large quantities (e.g., concrete, steel) are required for certain credits under the Envision and LEED rating systems. A number of free and fee-based LCA modeling tools exist. The Athena Institute offers ISO-compliant LCA-based tools in North America for whole buildings and assemblies. Athena Institute tools include a basic free EcoCalculator and a more detailed fee-based Impact Estimator for Buildings (Athena Institute, 2016). Other LCA software tools available at the time of this publication include GaBi (thinkstep, 2016) and SimaPro (SimaPro, 2016). Life cycle assessment modeling should ideally be performed early in the design process to support material selection decision-making. The end goal is to consider and select materials with the least effect on the environment and human health.

3.3.5.3   Sustainable Sourcing

A *sustainable source* is a manufacturer or supplier that is able to provide documentation to prove that its products (e.g., building materials) were, in fact, manufactured and delivered in a sustainable manner. The design team can specify submittals and product data to document compliance with project's quality and sustainability requirements. Example submittal information might include a copy of the manufacturer's sustainability policies, EPDs, and HPDs. There are also specific certifications available for select products such as textiles, hardwood, and plastic resins. Example certifications include Rainforest Alliance Certification, Certified Responsible Source, Environmentally Preferable Product. Certifications demonstrate via a third-party evaluator or audit that the product is, in fact, sustainably sourced. These types of certifications are rapidly evolving and vary by region and product type. Designers and owners are encouraged to investigate options for significant components and determine the extent of supporting documentation available. By requesting product certifications, or a corporate sustainability policy, as a submittal requirement during construction project owners are asserting marketplace demands on manufacturers and suppliers to pay more attention to their supply chain effects and corporate sustainability practices—particularly if these documents have not yet been developed.

3.3.5.4   End of Life Considerations

Designing facilities for reuse or recycling of materials and components, rather than disposal, is another element of sustainability. To facilitate reuse and recycling opportunities during future renovations, the design team should design for disassembly and deconstruction by maximizing mechanical fastening techniques and limiting use of adhesives. Where possible, provide layouts that allow for upsized piping and equipment and minimize the need for demolition and reconstruction of piping and equipment supports and pads.

It can be difficult finding uses for equipment with remaining useful life (e.g., standby generators that are being replaced due to larger load demand, variable-frequency drives, pumps, instrumentation, etc.). This equipment is typically salvaged to the contractor with the ultimate fate of equipment unknown. Depending on owner policies, an alternative approach may be to donate or hold a limited auction to transfer such equipment to another government agency, vocational wastewater training program, or community nonprofit group. This is an example of "up cycling" equipment rather than generating waste.

3.3.6 Future Proofing

Future proofing refers to the approach in which provisions are included to safeguard against future change scenarios such as regulatory change, climate change, and the evolution of technology.

3.3.6.1   Regulatory Change

Regulatory change is typically anticipated (e.g., toward higher quality effluent and biosolids), but the timing of such changes can be more difficult to predict. Open and frequent communication with WRRF regulators is encouraged to estimate this. When the risk of regulatory change within the life of the new facility or system is suspected, technologies should be selected that are reasonably easy to adapt to the expected future requirements. This does not suggest, however, that pumps and blowers should be over-sized as that approach reduces energy efficiency. Rather, adjust the expected service life for smaller equipment and provide space for adding or upsizing equipment later, when the trigger actually occurs.

3.3.6.2   Resilience to Climate Change

Resilience to climate change should be built into new projects and retrofit decisions. More extreme weather events, increased peaking factors, higher temperatures or greater swings in temperature, rising sea level, and wider flood plain areas are example considerations. In drought-prone areas, influent wastewater may get more concentrated and the value of reclaimed water will likely rise. The sensitivity of the design should be checked to see how resilient the WRRF will be to potential climate-related changes.

3.3.6.3   Technology Migration Pathways

It is important to provide flexibility in the layout, tankage, hydraulic profile, and equipment access to accommodate future technologies. Providing a technology migration pathway extends the useful life of the facility, increases resiliency and adaptability to change, and should provide a financial advantage when viewed from a life cycle cost perspective. An example of this is to provide extra freeboard and a future hydraulic loss to add lamella plates, membranes, or suspended media into the secondary treatment process. Migrating from simple sidestream aeration to an anammox process or planning connections and space for a future phosphorus-recovery system are other examples.

3.3.7 Other Sustainability Considerations During Final Design

Designers are encouraged to pay attention to the high-energy demand equipment. With electricity being the largest type of energy consumed, the sizing and selection of high-load electrical equipment such as pumps, blowers, and centrifuges should be scrutinized.

- Efficiency at lower flows—The sizing and selection of high-load electrical equipment (e.g., pumps, blowers, centrifuges, UV, dryers) should be scrutinized across the range of flows and loads. Energy efficiency should be sought across this range, not just at the design peak flow in the 20-year future period. Check pump and blower curves for operating conditions in the first year of operation and consider more units of smaller capacity to improve efficiency at low and average design operating points. Demonstrating energy efficiency at startup is a common condition of receipt of grant and loan funds. U.S. EPA offers guidance to states regarding efficiency requirements for Clean Water State Revolving Fund–funded projects. Modular systems, which add capacity in discrete increments, have many advantages.

- Control strategies—Consider options to provide a high degree of system control by using control strategies that minimize energy consumption (e.g., automatic dissolved oxygen control within zones, use of most open valve technology; control blower output based on total flow). Evaluate if flow attenuation/equalization could reduce the range of operation and improve certainty to support selection of smaller motor sizes and more efficient equipment.

- Maintainability—Providing systems such that preventive and routine maintenance can easily be performed is more likely to result in a properly maintained system operating per design intent. Engage maintenance staff during the design phase to get input on access around equipment, preferences for standardized components, use of modular systems, in-place cleaning, and other features that improve ease of maintenance.

During final design, consider the final details or specification language that could be included, at zero/low marginal cost that would improve sustainability of the construction phase and/or the resulting installation. Examples include the following.

- Energy performance specifications—Specifications that emphasize energy performance and detail testing and commissioning steps for the contractor are recommended. Project documents can define the required factory witness testing and modeling of energy consumption for high-energy demand equipment; establish maximum power draw at measureable design conditions; include protocols for measurement during field startup testing; define supplier's mitigation options and timelines for corrective action; and set clear and justifiable penalty based on life cycle cost for non-compliance.

- Hydraulic profile—Attention should be paid to the hydraulic profile and elevation drop between tanks, so that the forward wastewater flow is only pumped once wherever possible.

- Automation and monitoring—Building automation systems (e.g., EMIS) to monitor and control temperature, humidity, measure and report flows of energy and water, and track other parameters in facility buildings.

- Heating systems—Verify hot water heating systems can be interconnected (for reliability, redundancy); include thermostats that are programmable and include outside air-turndown when areas are unoccupied.

- Sustainable construction practices—This would include local sourcing, construction and demolition waste diversion, use of recycled materials, noise controls, minimizing visual and mobility effects, and air quality considerations.

- Noise—Specify allowable noise levels clearly; require noise attenuation; include noise measurement during startup testing; set penalty for non-compliance.

- Landscaping—Include sustainable landscapes (quick establishment, no irrigation once established, pesticide and fertilizer free); specify native plantings; use biosolids and reclaimed water; include shade trees; plan to manage stormwater via green stormwater infrastructure (GSI).

- Lighting—Specify energy-efficient lighting and lighting controls, such as motion sensors, timers, and photo cells.

- Pavement—Specify use of recycled pavements or permeable pavements; GSI; specify low heat-island materials (e.g., solar reflective index, SRI, of 29 or less).

3.4   Bidding Phase

The bidding and procurement phase for sustainable and energy management projects may be comparable to conventional construction or it may proceed using an alternative procurement approach and/or financing arrangement. This section identifies possible project financing and implementation options for energy-efficiency improvement type projects and offers general tips about how to alert the construction community during competitive bidding of unique project requirements.

3.4.1 Financing Energy and Sustainability Projects

Sustainable project funding options include

- Energy Performance Contracting—Energy performance contacting (EPC) offers "turnkey solutions" that provide customers with a comprehensive set of energy efficiency, renewable energy, and distributed generation measures. Energy performance contacting is accompanied with guarantees that the savings produced by a project will be sufficient to finance the full cost of the project. A typical EPC project is delivered by an energy services company (ESCO), or qualified contractor who provides a broad range of energy solutions including designs and implementation of energy savings projects, retrofitting, energy conservation, energy infrastructure outsourcing, power generation, and energy supply. Because there are financing fees associated with this approach, EPC may be most advantageous when the facility owner does not have access to sufficient upfront capital. An EPC generally consists of the following elements:
  - Turnkey Service—The ESCO or qualified contractor provides all of the services required to design and implement a comprehensive project at the customer facility, from the initial energy audit through long-term monitoring and verification of project savings;
  - Comprehensive Measures—The ESCO or qualified contractor tailors a comprehensive set of measures to fit the needs of a particular facility, and can include energy efficiency, renewables, distributed generation, water conservation, and sustainable materials and operations;
  - Project Financing—The ESCO or qualified contractor arranges for long-term project financing that is provided by a third-party financing company. Financing is typically in the form of an operating lease or municipal lease; and
  - Project Savings Guarantee—The ESCO or qualified contractor provides a guarantee that the savings produced by the project will be sufficient to cover the cost of project financing for the life of the project.
- Utility energy service contracting—This is similar to an EPC but typically refers to contracts between agencies within the federal government and a local power or natural gas utility.

- Power purchase agreements—A power purchase agreement is a contract between two parties, one that generates electricity (the seller, e.g., the WRRF) and one that is looking to purchase electricity (the buyer, e.g., nearby industry).

- Grants and rebates—Local energy companies may offer efficiency incentive programs in the form of grants and rebates. States may also offer special funding programs for energy-efficiency related upgrades.

- Privatization—Some WRRFs have privatized portions of their facility, allowing a private company to own and/or operate a discrete system or unit process (e.g., CHP, biomethane recovery) under a contractual relationship. Typically, the private party pays the upfront capital cost of the installation located within the facility perimeter.

- Joint ownership—Joint ownership is similar to privatization, but with a defined ownership (e.g., 50:50) in the constructed works by both parties. The parties may be municipalities, corporations, nonprofits, or private entities.

- Shared savings plans—These plans are another form of agreements between WRRFs and the local energy utility. Shared savings plans include a mechanism that adjusts the loss in revenue to the utility that resulted from energy conservation efforts undertaken by the WRRF in exchange for upfront capital or other incentive offered by the utility to the WRRF to support project implementation.

- Performance contracting—The energy efficiency goals (e.g., kWh/MG targets) and effluent quality goals are stated in the contract documents and include both rewards for performance or betterment and penalties for failing to perform.

- Biofuels credits—Under the Renewable Fuel Standard program (United States only), the sale of biomethane, ethanol, and biodiesel derived from WRRFs is also eligible for supplemental revenue from the sale of Renewable Identification numbers (RINs). Sale of RINs is usually brokered by a third party. The Wastewater Reclamation Authority of Des Moines, Iowa, is an example WRRF that will be injecting biomethane into the natural gas utility pipeline and also receiving RIN revenue once construction is complete (expected 2018). See https://www.epa.gov/renewable-fuel-standard-program for more information.

Be sure to evaluate local regulations to verify feasibility for any of the aforementioned financing approaches. When outside financing will be sought, it is also important to start working with outside partners, regulators, and local energy efficiency incentive providers as early as possible.

3.4.2 Emphasizing Sustainability and Energy Management for Competitive Bid Procurement

To emphasize the importance of a project's sustainability and energy management goals on prospective general contractors during the bidding phase, consider the following:

- Confirm the general conditions listed in the contract documents align with sustainability and energy management goals before bid advertisement;
- Conduct pre-bid outreach to general contractors and subcontractors recognized for sustainable/green building achievements;
- Contract documents should have specific assignment of responsibility for intended sustainability elements:
  - Personnel responsible for gathering and submitting LEED or Envision documentation;
  - Commissioning agents and third-party reviewers;
  - Require performance guarantees regarding energy consumption and GHG emissions, as well as treatment performance;
  - Consider configuring the bid form to award to lowest life cycle cost (not just lowest capital cost) to allow for consideration of O&M expenses;
  - Evaluate boilerplate requirements on testing/commissioning, performance guarantees, and liquidated damages in relation to sustainability goals. Discuss potential to strengthen this language to hold contractors/vendors accountable for energy and sustainability requirements (e.g., include in schedule of values); and
  - Clear requirements for measuring energy performance at startup and for compliance checks of other project sustainability requirements (e.g., sustainable sourcing submittals), including links to standard conditions that define key terms (e.g., "Acceptance") and/or set financial penalties for failing to comply.
- Emphasize sustainability features at pre-bid meeting.

A useful reference for RFP guidance is available from the Consortium for Energy Efficiency (2010). The Consortium for Energy Efficiency (CEE) offers guidance on issuing RFP/RFQ documents that seek to include for energy efficiency elements into a new or refurbished WRRF (CEE, 2010).

3.5   Construction Phase

The contractor and owner's construction management (CM) staff are integral players in ensuring facilities are built in the intended sustainable manner.

Through design involvement resulting from the ID process, CM staff should have a complete understanding of the sustainable project components and how these components are to be implemented via the contract plans and specifications (contract documents). Attention to sustainability and energy management goals and objectives continues into the construction phase.

There are two primary considerations related to environmental management construction practices—construction practices affecting sustainable project delivery and construction contract execution. Sustainable construction practices relate to how the contractor executes the work, including means and methods. Construction contract execution in this context refers mainly to the submittal process. The contract documents must cover both.

The following are examples of sustainable construction practices:

- Construction constraints and milestones to minimize community effects—phasing work to minimize traffic effects or visual effects (unsightly construction site);
- Construction fencing to minimize visual effect on the community;
- Onsite reuse of excavated materials;
- Air monitoring to protect the community or workers from diesel fumes, dust, or other airborne contaminants;
- Sound curtains to minimize noise;
- Dust control to protect electrical equipment;
- Heating, ventilation, and air conditioning filter monitoring and changeout to protect the workspace and new equipment;
- Construction and demolition waste requirements to divert landfill waste and better control lower level hazardous waste disposal, such as fluorescent lamps or mercury switches;
- Equipment idling times and type of backup warning alarm (to minimize nuisance fumes and noise); and
- Construction management trailer/workspace with sustainable features, such as increased insulation, water efficient fixtures, recycled content, programmable thermostat, LED lights, EnergyStar equipment.

Construction contract execution requires adherence to the contract documents, including the sustainability requirements integrated into these legal documents. Clear specification language and submittal requirements are important to ensuring the contractor delivers the intended sustainability components and implements the project in a sustainable manner.

Depending on the complexity of sustainability requirements and whether the project is seeking a verified sustainability award, a specific sustainability submittal/credit tracking system or responsible staff person may be needed.

Some sustainability components are relatively easy to specify and confirm (e.g., motor control centers must have power monitoring capability). Other elements, such as those related to contractor methods and means, may be more difficult to enforce (e.g., equipment idling times during construction). It may also be unclear during the design phase where recycled materials may be found or whether on-site excavated materials will be suitable for re-use with sufficient confidence to include stringent requirements in the specifications. One strategy to address this issue and incentivize the contractor to focus on sustainability is to establish an upfront change-order budget set-aside (e.g., for greater use of recycled materials and/or reducing emissions during construction) and provide for an early collaboration workshop between the owner, designer, and contractor post award to evaluate options post award.

3.6   Operations and Maintenance Phase

Sustainability planning, engineering, financing, and construction efforts are lost if the new facilities are not operated or maintained as intended. Operations and maintenance staff should be engaged early in the design via the ID process so they understand and support sustainability-related project elements, and operate and maintain those elements as intended. It is imperative for the design team to listen to and address O&M constraints and concerns expressed about sustainability elements and operational changes. When O&M staff understand why sustainable project elements were included, how those elements fit into the agency's goals or requirements, and receive proper training, there is a much higher likelihood that the sustainable elements will be successfully incorporated to facility operations, and that those elements will be properly maintained.

To preserve the functionality of new sustainable elements and to best inform new personnel, the design team should consider providing a transfer document to include with the O&M manuals. The transfer document summarizes sustainable elements associated with the facility so that O&M staff can incorporate methodologies into SOPs.

Sustainable elements considerations with O&M in mind include the following:

- Consider incentives to share savings—The Lowell Regional Wastewater Utility (LRWWU) (Lowell, Massachusetts) instituted an incentive program whereby employees are given financial rewards from the city manager for successful third-party audits and maintaining certification. In addition to many non-energy-related environmental improvements like reduced odor complaints and a robust recycling program, the LRWWU facility has reduced energy use by more than 35% since 1995, with total savings estimated at more than $3 million. Projects implemented under the EMS include pump, blower, lighting, and HVAC efficiency upgrades; rooftop solar; and solar thermal heating;

- Energy monitoring systems—Identify example software systems for intelligent management of real-time data and analytics;

- Programmable thermostats to ensure heaters don't stay turned up when staff leaves the area;

- Lighting panel controls to allow staff to turn out all lights from a single location;

- Develop operating strategies and equipment settings and maintenance with O&M staff management and line staff to ensure a common understanding of sustainability targets throughout the chain of command;

- Ensure O&M manuals/SOPs/checklists include pertinent sustainability information, e.g., LED lamps, HVAC filters, temperature settings;

- Landscape items—Explain the function and O&M procedures for rain gardens, green roof, stormwater biofiltration, porous pavement, and native plantings;

- Reuse water—Consider effects on facility processes; provide space for supplemental filtration if needed;

- Integrated pest management—Consider this when specifying materials; integrate into SOPs; and

- Green maintenance and cleaning practices—O&M manuals should identify green cleaning products that can be used on new equipment and surfaces where possible.

4.0 REFERENCES

ASTM International (2017a) https://www.astm.org/COMMIT/sustain.html (accessed June 2, 2017).

ASTM International (2017b) https://www.astm.org/Standards/sustainability-standards.html (accessed June 2, 2017).

Athena Institute (2016) "Ecocalculator" online tool. http://www.athenasmi.org/our-software-data/ecocalculator/ (accessed May 9, 2017).

Building Research Establishment (2017) Building Research Establishment's Environmental Assessment Method (Breeam®). http://www.breeam.com/infrastructure (accessed June 2, 2017).

California Energy Commission (2011) *Water and Wastewater Energy Efficiency*; Publication CEC-500-2010-048. http://www.energy.ca.gov (accessed May 9, 2017).

Consortium for Energy Efficiency; Water Environment Federation (2010) *Energy Efficiency RFP Guidance for Water-Wastewater Facilities*. https://library.cee1.org/content/energy-efficiency-rfp-guidance-water-wastewater-facilities (accessed May 9, 2017).

Day, D. (2013) We Are Environmentalists. *Treatment Plant Operator Magazine*, February. http://www.tpomag.com/editorial/2013/02/we_are_environmentalists (accessed May 9, 2017).

Electric Power Research Institute (2013) *Electricity Use and Management in the Municipal Water Supply and Wastewater Industries*; Electric Power Research Institute: Palo Alto, California. http://www.epri.com/abstracts/Pages/ProductAbstract.aspx?ProductId=000000003002001433 (accessed May 9, 2017).

EnviroSim (2016) BIOWIN Software V4.1.

Fitzpatrick, J.; George, R.; Caliskaner, O. (2015) Paper presented at LIFT: Innovative Advanced Primary Treatment for the Utility of the Future; Mobile Session 231 at the 88th Annual Water Environment Federation Technical Exposition and Conference; Chicago, Illinois, Sep 28.

International Living Future Institute (2016) *The Red List*. https://living-future.org/declare/declare-about/red-list/ (accessed May 9, 2017).

International Organization for Standardization (2011) *ISO 50001:2011—Energy Management*. http://www.iso.org/iso/home/standards/management-standards/iso50001.htm (accessed June 2, 2017).

International Organization for Standardization (2015) *ISO 14001:2015 Environmental Management Systems—Requirements with Guidance for Use*. https://www.iso.org/iso-14001-environmental-management.html (accessed June 2, 2017).

International Organization for Standardization (2006a) *ISO 14040:2006 Environmental Management—Life Cycle Assessment—Principles and Framework*. https://www.iso.org/standard/37456.html (accessed June 2, 2017).

International Organization for Standardization (2006b) *ISO 14044:2006 Environmental Management—Life Cycle Assessment—Requirements and Guidelines*. https://www.iso.org/standard/37456.html (accessed June 2, 2017).

Institute for Sustainable Infrastructure (2017). https://sustainableinfrastructure.org/ (accessed June 2, 2017).

Maine Department of Environmental Protection (2010) Model Energy Audit RFP. https://www3.epa.gov/region9/waterinfrastructure/docs/model-energy-audit-request.pdf (accessed May 9, 2017).

New York State Energy Research and Development Authority (2016) Library of Water and Wastewater Technical Reports. https://www.nyserda.ny.gov/About/Publications/Research-and-Development-Technical-Reports/Water-and-Wastewater-Technical-Reports (accessed May 9, 2017).

Pacific Northwest Pollution Prevention Resource Center Home Page (2015) http://pprc.org/index.php/2015/p2-rapid/is-high-density-polyethylene-hdpe-a-good-choice-for-potable-water/ (accessed May 9, 2017).

Seymour, D. (2009) *Can Nutrient Recovery Be a Financially Sustainable Development Objective?* PNCWA, Sept 15, 2009. Boise, Idaho. http://www.pncwa.org/assets/documents/2009%20PNCWA-%20Session%2014-1%20-%20Centrate%20to%20Fertilizer%20-%20David%20Seymour.pdf (accessed May 9, 2017).

SimaPro (2016) SimaPro LCA online tool. https://simapro.com/ (accessed May 9, 2017).

Stoner, N. K. (2011) *Working in Partnership with States to Address Phosphorus and Nitrogen Pollution through Use of a Framework for State Nutrient Reductions*, Memorandum to EPA Regional Administrators, March 16.

thinkstep (2016) GaBi LCA online tool. http://www.gabi-software.com/america/index/ (accessed May 9, 2017).

U.S. Environmental Protection Agency (2005) *EMS Case Studies in the Public Water Sector.* http://ocw.unesco-ihe.org/pluginfile.php/3163/mod_resource/content/1/EMS-Water_Sector_-_USEPA_-2005.pdf (accessed May 9, 2017).

U.S. Environmental Protection Agency (2014) *Food Waste to Energy: How Six Water Resource Recovery Facilities are Boosting Biogas Production and the Bottom Line.* https://archive.epa.gov/region9/organics/web/pdf/epa-600-r-14-240-food-waste-to-energy.pdf (accessed May 9, 2017).

U.S. Environmental Protection Agency (2015) *Adaptation Strategies Guide*; EPA 817-K-15-001; Office of Water (4608-T). https://www.epa.gov/sites/production/files/2015-04/documents/updated_adaptation_strategies_guide_for_water_utilities.pdf (accessed May 9, 2017).

U.S. Environmental Protection Agency (2016) *ENERGY STAR Portfolio Manager, Wastewater Treatment.* https://www.energystar.gov/buildings (accessed May 9, 2017).

U.S. Environmental Protection Agency. https://www.epa.gov/crwu (accessed May 9, 2017).

U.S. Green Building Council (2016) *LEED V4 for Building Design and Construction Users Guide*; U. S. Green Building Council: Washington, D.C. http://www.usgbc.org/LEED/ (accessed June 2, 2017).

Water Environment & Reuse Foundation (2015) *Demonstrated Energy Neutrality Leadership: A Study of Five Champions of Change (ENER1C12b)*; Water Environment & Reuse Foundation: Alexandria, Virginia.

Water Environment Federation (2010) *Energy Conservation in Water and Wastewater Facilities*; Manual of Practice No. 32; Water Environment Federation: Alexandria, Virginia.

Water Environment Federation (2013) *The Energy Roadmap: A Water and Wastewater Utility Guide to More Sustainable Energy Management*; Water Environment Federation: Alexandria, Virginia.

Water Environment Federation; American Society of Civil Engineers; Environmental Water Resources Institute (2018) *Design of Water Resource Recovery Facilities*, 6th ed.; WEF Manual of Practice No. 8/ASCE Manuals and Reports on Engineering Practice No. 76; Water Environment Federation: Alexandria, Virginia.

Water Environment Research Foundation (2011) *Exploratory Team Report—Energy Management*; Water Environment Research Foundation: Alexandria, Virginia.

Water Environment Research Foundation (2014) *Utilities of the Future Energy Findings* (ENER6C13); Water Environment Research Foundation: Alexandria, Virginia.

Water Research Foundation (2012) Database of Best Practices for Energy Efficiency (compilation of case studies). http://www.waterrf.org/resources/pages/PublicWebTools-detail.aspx?ItemID=13 (accessed May 9, 2017).

Willis, J. L., Yuan, Z.; Murthy, S. (2016) Wastewater GHG Accounting Protocols as Compared to the State of GHG Science. *Water Environ. Res.*, 88, 704–714.

Young, M. (2016) Personal Communication with Jason Turgeon on October 12, 2016.

5.0   SUGGESTED READINGS

American Society of Civil Engineers (2013) *Infrastructure Reportcard*. http://www.infrastructurereportcard.org/a/#p/wastewater/conditions-and-capacity (accessed May 9, 2017).

American Society for Testing and Materials (2011) *Test Method for Determining Air Change in a Single Zone by Means of a Tracer Gas Dilution, E741.* http://www.astm.org/Standards/E741.htm (accessed May 9, 2017).

American Society for Testing and Materials (2016) *Sustainable Standards Development.* http://www.astm.org/COMMIT/sustain.html (accessed May 9, 2017).

American Society for Testing and Materials (2015) *D1598, Test Method for Time-to-Failure of Plastic Pipe Under Constant Internal Pressure.* http://www.astm.org/cgi-bin/resolver.cgi?D1598-15a (accessed May 9, 2017).

American Society for Testing and Materials (2015) *E1057, Practice for Measuring Internal Rate of Return and Adjusted Internal Rate of Return for Investments in Buildings and Building Systems.* http://www.astm.org/cgi-bin/resolver.cgi?E1057-15 (accessed May 9, 2017).

American Society for Testing and Materials (2015) *E1074, Practice for Measuring Net Benefits and Net Savings for Investments in Buildings and Building Systems.* http://www.astm.org/cgi-bin/resolver.cgi?E1074-15 (accessed May 9, 2017).

American Society for Testing and Materials (2010) *E779, Test Method for Determining Air Leakage Rate by Fan Pressurization.* http://www.astm.org/cgi-bin/resolver.cgi?E779-10 (accessed May 9, 2017).

American Society for Testing and Materials (2015) *E917, Practice for Measuring Life-Cycle Costs of Buildings and Building Systems.* http://www.astm.org/cgi-bin/resolver.cgi?E917-15 (accessed May 9, 2017).

American Society for Testing and Materials (2015) *E964, Practice for Measuring Benefit-to-Cost and Savings-to-Investment Ratios for Buildings and Building Systems.* http://www.astm.org/cgi-bin/resolver.cgi?E964-15 (accessed May 9, 2017).

American Society of Civil Engineers (2016) *Green Technologies for Sustainable Water Management*; American Society of Civil Engineers: Washington, D.C.

American Water Works Association (2014) *Current Sustainability Infrastructure Practices 2014.* http://www.awwa.org/Portals/0/files/resources/water%20utility%20management/CSIP%20Report%2012-30-14protected.pdf (accessed May 9, 2017).

American Water Works Association Research Foundation (Denver, Colorado) Home Page. http://www.awwarf.org (accessed May 9, 2017).

British Columbia Building Performance Study (2014) Rep. Vancouver: Lighthouse, Dec. 2014. Web. 6 Aug. 2016. http://www.sustainablebuildingcentre.com/wp-content/uploads/downloads/2014/12/BC-BUILDING-PERFORMANCE-STUDY_Dec-2014.pdf (accessed May 9, 2017).

California References: Climate Action Plan, Calculators, etc. http://www .coolcalifornia.org/article/climate-calculators (accessed May 9, 2017).

Energetics Incorporated (2015) *Energy-Positive Water Resource Recovery Facility Workshop Report.* http://www.energy.gov/eere/bioenergy/energy-positive-water-resource-recovery-workshop-report (accessed May 9, 2017).

Institute for Sustainable Infrastructure (2015) *ENVISION Rating System for Sustainable Infrastructure*; Institute for Sustainable Infrastructure: Washington, D.C.

Integrated Design Canada Roadmap, Natural Resources Canada URL: http://www.greenspacencr.org/events/IDProadmap.pdf (accessed May 9, 2017)

Life-Cycle-Based-Sustainability-Standards-Guidelines.pdf (ISO Docs) https:// www.pre-sustainability.com/download/Life-Cycle-Based-Sustainability-Standards-Guidelines.pdf (accessed May 10, 2017).

National Renewable Energy Laboratory (2009) *Handbook for Planning and Conducting Charrettes for High Performance Projects*, 2nd ed.; National Renewable Energy Laboratory: Golden, Colorado. http://www .nrel.gov/docs/fy09osti/44051.pdf (accessed May 9, 2017).

Nault, J.; Papa, F. (2015) *Lifecycle Assessment of a Water Distribution System Pump*; American Society of Civil Engineers: Toronto, Ontario. Lifecycle-Assessment-of-a-WDS-Pump-Nault-and-Papa-2015.pdf (accessed May 9, 2017).

New England Interstate Water Pollution Control Commission (2016) *TR-16, Guides for the Design of Wastewater Treatment Works;* NEISWPCC: Lowell, Massachusetts; (energy efficiency throughout and resiliency to climate added in revision; 2011 edition as revised in 2016).

New York State Energy Research & Development Authority (2010) *Water & Wastewater Energy Management Best Practices Handbook;* NYSERDA: New York; September. file:///C:/Users/strehlerjl/Downloads/NYSERDA-Water-Wastewater-Energy-Management-Best-Practices-Handbook.pdf (accessed June 2, 2017).

Operations and Maintenance Report—Multi-Agency Benchmarking Study, Dec 1999. In consultation with Damon S. Williams Associates, L.L.C. http://your.kingcounty.gov/dnrp/library/wastewater/wtd/pubs/ 9912Benchmarking/om.pdf (accessed May 9, 2017).

U.S. Department of Agriculture (2014) *Biogas Opportunities Roadmap.* http:// www.usda.gov/oce/reports/energy/Biogas_Opportunities_Roadmap_ 8-1-14.pdf (accessed May 9, 2017).

U.S. Environmental Protection Agency (2008) *Ensuring a Sustainable Future: An Energy Management Guidebook for Wastewater and Water*

Utilities. https://nepis.epa.gov/Exe/ZyPURL.cgi?Dockey=P1003Y1G.txt (accessed May 9, 2017).

U.S. Environmental Protection Agency (2010) *Evaluation of Energy Conservation Measures for Wastewater Treatment Facilities;* EPA 832-R-10-005; U.S. Environmental Protection Agency: Washington D.C.

U.S. Environmental Protection Agency (2011) *Opportunities for Combined Heat and Power at Wastewater Treatment Facilities: Market Analysis and Lessons from the Field.* https://www.epa.gov/chp/opportunities-combined-heat-and-power-wastewater-treatment-facilities-market-analysis-and (accessed May 9, 2017).

U.S. Environmental Protection Agency (2015) *Energy Efficiency in Water and Wastewater Facilities: A Guide to Developing and Implementing Greenhouse Gas Reduction Programs.* https://www.epa.gov/sites/production/files/2015-08/documents/wastewater-guide.pdf (accessed May 9, 2017).

U.S. Environmental Protection Agency (2016b) *GHG Inventory of Greenhouse Gas Emissions and Sinks 1990–2014;* EPA 430-R-16-002. https://www3.epa.gov/climatechange/Downloads/ghgemissions/US-GHG-Inventory-2016-Main-Text.pdf (accessed May 9, 2017).

U.S. Environmental Protection Agency (not dated) *Sustainable Management of Construction and Demolition Materials.* https://www.epa.gov/smm/sustainable-management-construction-and-demolition-materials (accessed May 9, 2017).

Water Environment Federation; National Biosolids Partnership; Water Environment Research Foundation (2013) *Enabling the Future: Advancing Resource Recovery from Biosolids;* Water Environment Federation: Alexandria, Virginia. http://www.wef.org/globalassets/assets-wef/3---resources/topics/a-n/biosolids/technical-resources/enabling-the-future.pdf (accessed June 2, 2017).

Water Environment & Reuse Foundation Home Page. http://www.werf.org (accessed May 2017). https://www.werf.org/a/ka/Search/Research Profile.aspx?ReportId=ENER1C12 (accessed May 9, 2017).

Water Environment & Reuse Foundation (not dated) *Intensification of Resource Recovery Report.* https://www.werf.org/a/ka/Search/Research Profile.aspx?ReportId=TIRR1R15 (accessed May 9, 2017).

Water Environment Research Foundation (2013) *Triple-Bottom Line Evaluation of Biosolids Management Options* (ENER1C12a); Water Environment Research Foundation: Alexandria, Virginia.

Water Environment Research Foundation (2015) *A Guide to Net Zero Energy Solutions for Water Resource Recovery*; Water Environment and Reuse Foundation: Alexandria, Virginia.

Index

CPSIA information can be obtained
at www.ICGtesting.com
Printed in the USA
LVHW051108150121
676332LV00001B/5

9 781572 783416